Lily
Pond

Also by Hope Ryden:

Books for Adult Readers

Bobcat Year
God's Dog
Mustangs: A Return to the Wild
America's Last Wild Horses

Books for Young Readers

Wild Animals of Africa ABC
Wild Animals of America ABC
The Beaver
America's Bald Eagle
Bobcat
The Little Deer of the Florida Keys
The Wild Pups: The True Story of a Coyote Family
The Wild Colt

Lily Pond

Four Years with a Family of Beavers

Text and Photographs by
HOPE RYDEN

Lyons & Burford, Publishers

Printed in the United States of America

10 9 8 7 6 5 4 3 2 1

Library of Congress Cataloging-in-Publication Data

Ryden, Hope.
    Lily Pond : four years with a family of beavers /text and photographs by Hope Ryden.
        p.   cm.
    Originally published: 1st ed. New York: W. Morrow, c 1989. With new afterword.
    Includes bibliographical references and index.
    ISBN 1-55821-455-0 (pb)
    1. Beavers. 2. Beavers—New York—Harriman State Park. 3. Ryden, Hope—Diaries. 4. Harriman State Park (N.Y.) I. Title.
[QL737.R632R94 1996]
599.32'32—dc20   96-28176
                                              CIP

Acknowledgments

I am grateful to New York State Park Superintendent Ken Krieser for granting me permission to make nighttime beaver observations in Bear Mountain and Harriman Mountain state parks. I also wish to thank John Mead, former curator of the Palisades Park Trailside Museum, for introducing me to an orphan kit he was raising. Special acknowledgment is owed Dr. Joseph Larson, professor of forestry and wildlife management at the University of Massachusetts in Amherst. Dr. Larson guided me to many recent studies and important sources of information on beavers. New York State Game Warden Kenneth Didion brought much reassurance into my life by keeping an eye out for people who would poach my subjects. My thanks. I feel particularly indebted to Hope and Cavit Buyukmihci (I regret to say that Cavit died while this book was in progress) for their hospitality at Unexpected Wildlife Refuge in New Jersey's Pine Barrens. (To join Hope's organization, The Beaver Defenders, see the address in the Appendix.) There aren't words to fully express my appreciation to writer-naturalist John Miller, who appears throughout this book. As readers will discover, John was of inestimable help to me throughout my four-year beaver study. I am also touched and grateful to neighbors and friends Dan, Henry, and Nina Pierson, and Tony Spadavecchia, who came to the aid of my colony during bad times. It was a pleasure to work with my enthusiastic editor, Liza Dawson. And finally, I wish to thank Dr. Donald Griffin for reading and critiquing this manuscript prior to its publication.

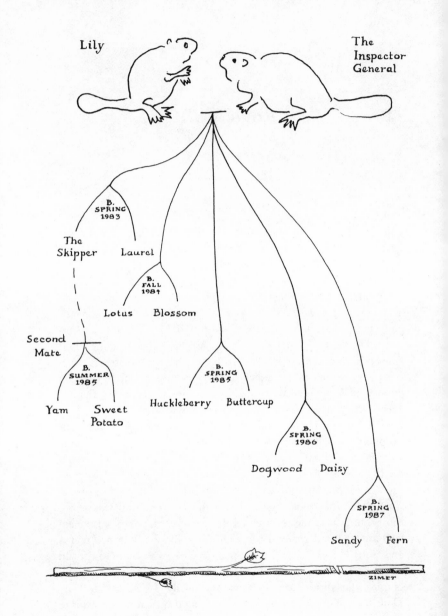

Lily

The
Inspector
General

B.
SPRING
1983

The
Skipper   Laurel

B.
FALL
1984

Lotus   Blossom

Second
Mate

B.
SUMMER
1985

Yam   Sweet
Potato

B.
SPRING
1985

Huckleberry   Buttercup

B.
SPRING
1986

Dogwood   Daisy

B.
SPRING
1987

Sandy   Fern

ZIMET

Preface
by Dr. Jane Goodall

I was entranced by *Lily Pond*. Not only is it a major contribution to our understanding of the natural history of beavers, but it points to the value of detailed observation of individually known animals over long periods of time.

The author's dedication and commitment kept her sitting quietly at the side of their pool night after night, regardless of her own comfort. The effort was worth it. She has painted a picture of beavers and their close family life that sparkles with the sheer joy she experienced as she became acquainted with them and began to understand how they communicated with each other—all the subtle sounds and gestures that enable them to get on with each other so well in the confined quarters of their lodge during the winter months when their world is covered in ice.

Hope Ryden appreciates the beauty and wonder of the natural world and she describes it powerfully. Her lively style, sense of humor, and unashamed love of being with and writing about the animals she came to know so well results in a book that will appeal to a wide audience. It will also benefit beavers since few people, once they have looked at these fascinating and intelligent creatures through Ryden's eyes, can fail to be captivated.

Reading this book was, for me, like journeying into a fascinating new world: I am enriched.

Introduction

Sometimes a new place gets such a grip on me, I have to wonder if I have been there before in some forgotten past life. Or perhaps the feeling is prescience, a hint of things to come, like a musical bridge that telegraphs the melody that must inevitably follow. It happened that way when I first saw Lily Pond. I fell under its spell and knew at once that it held some special meaning for me.

For several months I had been looking for a beaver pond to study, and not just any one would do. I had scouted active beaverworks in Vermont, spent cold nights at beaver-made ponds in Massachusetts, even stayed a week to photograph a colony of beavers in their self-styled habitat in the Pine Barrens of New Jersey; but none of these places was right. And so I continued to look, for if one plans to spend several hundred nights in a setting without consolation of human companionship, and weathering every condition throughout the seasons, one does not settle for the first site that promises to yield information. The place must also be compatible with one's being.

Then I saw Lily Pond, an impressionist painter's vision of what a pond ought to look like. Aquatic flowers of pristine perfection blanketed the water. On their round green leaves sat bug-eyed leopard frogs, and fat bumblebees buried their heads in the plant's floating blossoms. It happened to be morning and the place was busy with life.

As I edged along the rocky shore, I disturbed some painted turtles basking on a fallen tree. Plop, plop, plop, plop—one by one they leaped a distance ten times their height into the safety of the water. A beaver pond brews just the right nutrients to support a most complex tangle of life—a web of relationships that reaches onto land and stretches upward to sustain creatures of the air as well. So I looked skyward and there discovered cedar waxwings, swirling like fish, as they seined the blue air for the swarms of insects newly hatched off the water's surface. The trumpeting of a Canada goose brought my attention back to earth. The bird had seen me and was sounding an alarm signal to his mate and three young, grazing on the bank. Noiselessly, the family slipped into the pond and somehow managed to arrange themselves to look like a Venetian gondola—long-necked adult in front, long-necked adult in the rear, and five short-necked "passenger" goslings gliding in between.

I climbed onto the mud-and-stick dam that shored up this enchanting pond and hoped the beavers who had created it were still in residence and maintaining their engineering feat. For were they to abandon the site, the five-foot-high dam upon which I stood would gradually crumble under the pressure of the half million gallons of water it impounded, the pond would drain, and all the loveliness I was admiring would vanish.

"Don't set your heart on finding beavers here," John warned in a quiet voice. "I don't see enough food left around this pond to support a colony."

John, a longtime observer of beavers in Massachusetts, where he lived, had witnessed the decline of a number of ponds after the beavers who created them consumed all the resources necessary for their survival and departed. Nevertheless, I argued for beavers being present.

"What about all those swamp maples?" I asked, pointing to the south side of the pond. At first glance the steep embankment appeared a solid mass of mountain laurel, each twisted shrub of which was laden with pink and white blossoms. But here and there amid this floral luxuriance stood a silvery tree.

John did not bother to argue the point. He knew I had no wish to hear him reiterate my own case against the long list of authors who cite swamp maple as a significant beaver food. Already my observations had persuaded me that in some places, at least, beavers pass up this tree, for I had often found untouched stands at abandoned ponds.

"Ah, but this would be such a glorious place to observe a beaver colony," I mused, as I watched a blue kingfisher detach itself from a piece of matching sky, dive into the water, and flutter up to a tree snag with a minnow in its beak. When I looked around, however, I found I was talking to myself. John had begun searching for beaver sign along the north shore of the pond.

But the trees on that bank looked hardly better beaver food than did the waxen-leafed, fibrous-stemmed laurel that grew in such profusion across the way. The terrain John was exploring was relatively open and rocky and had the haunted look of a long-abandoned pasture, which indeed it was. Here and there stood a gnarled and oversized oak. Hardly worth a beaver's trouble to cut those aging specimens, I thought, for their crowns contained numerous dead limbs. Closer to the water more swamp maples clutched at the hard and rocky ground with their knucklelike roots exposed.

While John thrashed through a tangle of prickly barberry bush, which was fast colonizing the shoreline, I examined the muddy crest of the dam for beaver paw prints.

"Otters!" I called, upon discovering a five-toed hind track.

The place was looking more and more appealing. Observing animals is a slow and often tedious process. Under ideal conditions, a beaver-watcher can expect to obtain but a penny's worth of information for hours of time invested. During the long lulls, the presence of other kinds of animals can provide welcome diversion. And the otter, being the most playful of species, would be great fun to watch.

But there were still other reasons, in addition to the fact that the pond was so lovely and full of life, why I hoped to find a beaver family living in it. It just happened to be situated in Harriman State Park, a few minutes drive from my weekend cabin on Pothat Lake in New York's Ramapo Mountains. If beavers inhabited Lily Pond, I would not have to drop everything, leave daily life and friends behind, and take off for parts unknown on yet another extended field trip. I had had enough of the gypsy life during the years I had studied and written about wild horses, coyotes, bobcats, key deer, and bald eagles. As much as I treasured the memory of those experiences, I was ready to settle down. If this pond harbored an active beaver colony, I could live at home with my dogs during the day, drive to "work" in the evening (beavers are nocturnal), and return sometime after midnight to sleep in my own bed; in short, live a relatively normal life and have my adventure too.

Still another important consideration was the size and configuration of the pond—some four acres in the shape of an oversized football field. There were few irregularities along its shoreline—no coves or inlets or bends into which my prospective subjects could swim and thus elude detection. And the big stick-lodge, which I hoped was at that moment housing sleeping beavers, was ideally situated. Though a considerable distance from the dam on which I stood, it jutted into the water midway along the south bank. Were I to station myself directly across from it on the north shore, I could enjoy an unobstructed view of the comings and goings of all its tenants. Moreover, from that same observation point, I would be able to glass both ends of the pond with my binoculars. It would be like having a fifty-yard-line seat at my own beaver bowl.

Certain critical questions, of course, remained to be answered. Were beavers still living in a pond where food resources appeared so marginal, or had they already abandoned it? And if the lodge was occupied, did it contain more than a single animal? A bachelor beaver would not provide the kind of information I was seeking, since my main point in watching the species was to learn something about its social life.

"We might find some poplar branches and leave them on the dam, then come back tomorrow morning to see if they've been peeled," John suggested.

That seemed a practical idea. Anyway, it was morning and there was no use in our waiting for animals to appear. Beavers, as a rule, do not emerge from their lodge until early evening. Meanwhile there were other ponds to investigate in this state park.

Harriman State Park abuts Bear Mountain State Park, and together they encompass eighty-one square miles of Rockland County, New York. These siamese-twin parks are wild and rocky and high. Bear Mountain Park contains a geological formation called the Hudson Highlands, comprising many summits and tornes and scenic

The lodge, situated midway along the south shore, is all but unapproachable. Fronted by water, its backside is protected by dense mountain laurel.

overlooks along the Hudson River, and these extend westward into Harriman Park, where they collide with the Ramapo Mountain Range. This heavily forested region is laced with boulder-strewn, hemlock-shaded streams, and pocked with swamps and ponds and lakes, any one of which might contain beavers. Deer and fox and river otter and coyotes and raccoons and possums are residents of the area, as are a variety of smaller critters—skunks and mink and weasels and deer mice and shrews. And, of course, such wild and undeveloped spaces serve a great many people. The terrain in the two parks is hard-going, and provides a challenge to rock climbers and a wilderness experience for nature-lovers who, on summer weekends, stream in by car and bus from New York City just forty miles away. The twin parks are large enough to absorb this human onslaught and at the same time offer individuals in search of solitude a surprising degree of isolation; for the lay of the land is many-faceted, with the various mountain chains intersecting at odd angles. The geography of the place, in fact, can present a confusing picture to hikers who stray off blazed trails, and on occasion may even grant an adventure-seeker the heart-pounding thrill of being lost for a while. Imagine being lost in the woods only forty miles from New York City!

At night Harriman Park and Bear Mountain Park are returned to the animals. Camping is allowed, but only in designated areas and by special permit. Since I would need such permission to remain at a beaver pond after sundown, I was in a quandary. I did not want to apply for and be granted a permit to work nights at Lily Pond before ascertaining that beavers were actually there. By the same token, in order to find an active beaver colony anywhere at all in the park system might necessitate spending several nights at a number of different locations.

While deliberating about what move to make next, John and I located a stand of aspen and returned to Lily Pond with a bundle of branches. These we carried to a muddy stretch of shoreline that looked as if some animal had been using it as a docking point.

"If there are beavers here, I wonder how they will react to this windfall," John commented. "From the look of this place, they won't have tasted anything like this before."

Aspen is decidedly the preferred food of *Castor canadensis* and, like John, I was curious to know how beavers who had no familiarity with this tree would react to it. Would they know at once that it was good eating? Or might they pass it up? This was just another of the many ready-made experiments I frequently find waiting to be tested on the animals I observe. If our "bait" generated tracks, but remained untouched, I would have to suspect that learning and/ or imitation plays a significant role in the species' food selection. On the other hand, should I return to find the aspen debarked, and tooth-etched sticks strewn about the place, I would not only have confirmation that beavers inhabited the place, but I would also have acquired a bit of evidence that the species' love of aspen is inherent.

As the two of us sat by the pond, placing branches stem first in the water, I suddenly heard myself announce:

"Let's drive to park headquarters before the office closes, so I can get a permit to watch beavers here at Lily Pond."

John looked surprised. "You don't even know if there is a single beaver in this pond."

"That may be true, but I don't want to delay any longer. Besides, I feel certain this will be the place I will work. I can't explain why."

John stared at me with an expression of incredulity and started to say something, then changed his mind. I knew only too well what he was thinking. I had put him to a lot of trouble, prescouting ponds for me in Massachusetts, and I had driven the 200-mile distance from New York, where I live, to inspect them. Together we had located one that seemed exactly right. It was a classic beaverwork with upper and lower auxiliary pools, a six-foot-high dam and plenty of food to guarantee that the beavers there would remain in place for a while. But now my heart was set on Lily Pond, and John was having difficulty adjusting to this abrupt shift in my thinking. Still, being a nature writer himself, he knew perfectly well what it was like to fall in love with a wild place. He also understood what I would soon be up against, trying to watch, identify, and make sense out of the lives of hard-to-see beavers— animals that stay in their lodge during daylight hours and whose nighttime activities often take place underwater. The task would prove formidable, no matter where I tried it.

And so we headed for park headquarters.

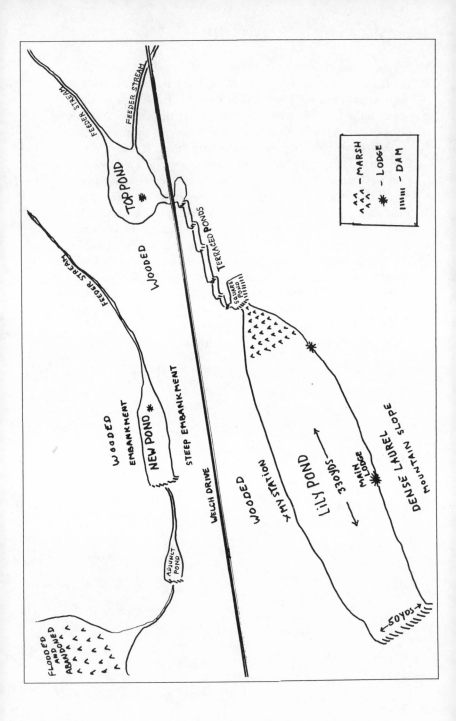

Chapter One

"There could be a dozen beavers living in that lodge over there, and we wouldn't see a single one," I said, breaking a long silence. "Even if they did come out, the lilies here are just too thick for us to see anything."

Fragrant water lilies and yellow bullheads blanketed the surface of the water in such matted profusion that the slightest breeze caused their overlapping pads to tip on edge, revealing some bright red undersides. The flapping leaves were distracting. They seized my attention again and again, as I scanned the pond for animal movement. But here and there were breaks in the floral mats, narrow bands of open water laid out in a geometric pattern to suggest that beavers had created them. One sparkling ring encircled the pond and several spokes extended from the bank lodge to various shore points.

This pattern gave me hope. Beavers make such swimming channels. They must in order to move about and transport the lumber they harvest to where it is needed. To have built the lodge we were looking at, for example, a huge conical affair made entirely of branches and mortared with mud, would have required a great many such deliveries. Trees had to be felled, then cut into manageable sections and dragged to the pond, where they could be launched and towed to the south shore. When possible, beavers create canals connecting harvestable tree stands to their ponds, for floating heavy lumber on water is considerably easier than dragging it across rough and broken ground. Getting materials to the pond does not, however, always solve a beaver's transport problems. Towing twiggy limbs across shallow places or through a solid mat of lilies is no easy trick. So beavers dredge deep channels on the pond bottom over which they can float their materials. At most ponds these underwater troughs are not apparent to an onlooker. Only after an abandoned pond has completely drained does one see the interlocking travel routes the colony created. Lily Pond, however, was an exception. Here the location of the beavers' shipping lanes could be deduced and even mapped by the absence of lilies growing in them. It was on these bands of open water that I focused my attention in the hope of spotting the wake of a swimming beaver.

How extraordinary, I thought, that a nonhuman animal could or would create canals and channels. At the end of the last century, when Giovanni Schiaparelli announced his discovery of "canals" on Mars, astronomers of the day were quick to interpret the finding as evidence of the presence of "something like our own race" on that planet. Were those nineteenth-century star gazers unaware that beavers here on earth perform the very feat that so excited them? Moreover, *Castor canadensis* does so superbly well. One beautifully engineered beaver canal in Colorado measures 750 feet long. Yet, paradoxically, a good many scientists who willingly entertain the possibility that "intelligence" exists elsewhere in the universe, are reluctant to see any sign of it in animals here on earth. The feats of subhuman creatures are downplayed. Beaver canals and channels and dams are dismissed as the end product of "instinctive behavior"—whatever that may mean.

Yet there is something truly remarkable about these engineering feats. The very idea that an animal would work hard at one task, that of dredging a canal or a channel, in order to lighten the labor of a subsequent task, that of moving materials from one site to another, is mind boggling. It prompts one to wonder if the species possesses foresight.

The permit I had obtained from the park superintendent, as it happened, was broadly worded. It allowed me nighttime access to any beaverworks I might find either in Harriman or in Bear Mountain Park. During the daylight hours, John and I checked out a number of possible sites, some of which had been suggested by park officials. But though we discovered evidence of past beaver

A beaver-dredged canal used to float harvested trees to the pond.

occupation in several places, we found no sign of current habitation, no freshly cut stumps or cropped sedge grass, no beaver-scent mounds or peeled sticks.

Beaver-made ponds appear and disappear cyclically, and the places we inspected were of great interest to both of us. They looked like beaver ghost-towns and appeared to have been occupied and abandoned more than once. That is the way of the species. *Castor canadensis* is nature's agent for renewal; the creature's appearance and disappearance create drastic alterations in a place, forcing old tired systems to yield to entirely different complements of plant and animal life. A new beaver pond serves a broad spectrum of aquatic species—otters, muskrat, mink, ducks, fish, turtles, frogs, wading birds—and continues to do so for as long as it is surrounded by a substantial number of the beaver's preferred food trees. When these are used up, the colony moves on, its forsaken dams break, and the pond drains. Then the rich mucky bottoms of what once were beaver waterworks give rise to an entirely different type of vegetation. Meadow plants take root and grow and these support a new array of animals, deer and voles and rabbits, which, in turn, become the food base for land predators, foxes, bobcats, coyotes, weasels, hawks. It is not difficult for a practiced eye to recognize that a particular meadow arose from the fertile sediment left by repeated beaver habitation. But as time passes, this lush meadow gives way to yet another biotic system. It is colonized by trees, the first to pioneer being willow, birch, and aspen—species the beaver relishes. As the forest matures, beavers once again return and turn the place into a pond. First they dam what water trickles through the wooded tract, thus drowning a certain number of trees. To the human eye, this looks like an act of vandalism. Whitened and crownless trunks stand like ancient columns of a Roman ruin, lifeless monuments to a dead past. But in reality, the situation is quite different. Even in this moribund state, the trees' rotting cores serve another succession of animals, whose turn it is to thrive. Woodpeckers, owls, kingbirds, and flying squirrels find nesting sites in their decaying trunks. Nuthatches, chickadees, and brown creepers feed on the insect life that proliferates in the rotting wood. Great blue herons construct huge nests on the forked tops of these forest relics. The big gangling birds, whose eight-foot wingspans prohibit flight through dense canopies, now enjoy plenty of clearance to take off and land on the towering eyries they construct. And at the base of these huge nests, crayfish and other aquatic creatures breed in the rising water, providing the herons plenty of food for their chicks. But this phase also passes. When the rotting trunks of the drowned trees become so weakened by decay that they topple into the water, stored nitrogen is released and settles on the pond bottom, enriching it for that future day when the site will once again explode with meadow plants.

This biotic succession, generated by the beaver, serves many functions. It prevents soil from becoming exhausted through overuse by any single "crop" and allows different forms of flora and fauna to have their day. It is nature's rest-and-rotation plan, a system widely imitated by modern-day agriculturists.

But, I wondered, at what stage in this ever-changing cycle was Lily Pond? The absence of any of the beaver's preferred food trees, namely aspen, willow, birch, and alder, suggested that its aquatic phase might be on the wane. On the other hand, perhaps there were food resources here that I failed to recognize. Little is known of the beaver's use of aquatic plants, and a wide variety of these species were in bloom.

One encouraging sign was the condition of the 150-foot-long dam. It looked in good repair. Obviously, the work had been in place for some years, for sections of its crest supported mature vegetation. Even a few trees had taken root on it; one swamp maple had grown to a considerable size. And judging from tracks along its crest, deer habitually used it as a bridge; over many years' time their trampling hooves had packed and reinforced the structure. It was upon this engineering feat that John and I were perched when he suddenly pointed out that the water directly in front of us was gently rocking.

"I think it's a beaver," he said with deliberate restraint in his voice, not wanting to be party to any enthusiastic outburst such a statement might provoke in me.

Just then a furry face peered out of the water. Then slowly surfaced one long brown body, to which was attached a large, hairless, paddle-shaped tail. After idling for a few moments, the animal glided past us, riding low in the water like a barge carrying too much cargo.

"That's one big beaver," I gasped.

The creature seemed unaware of us, as we remained frozen in position. Like all wild animals, beavers are keenly alert to any movement that flickers across their field of vision, while at the same time they may fail to see a large and motionless human being directly in front of them. Twice the animal paddled the length of the dam and back again, allowing us close-up views of him. On his third pass, he dived and covered the same course again, but this time underwater and trailing bubbles.

I seized the opportunity to shift into a more comfortable position and whispered to John that the beaver seemed to be inspecting the dam for leaks. He nodded in agreement, then glanced at me and we both laughed.

"Have you got a name for him yet?" John asked, as he flung a lanky arm around me in a congratulatory gesture.

"Well, for now he'll have to be Beaver Number One. Let's hope there will be a Beaver Number Two and a Beaver Number Three and a Beaver Number Four. Let's hope he doesn't turn out to be a confirmed bachelor," I answered in a controlled voice, trying hard to conceal the excitement I was feeling.

We watched Beaver Number One surface at the far end of the dam and then, through binoculars, tracked his wake as he slowly moved along a spoke of open water, a channel that connected dam and lodge. There he dived and presumably entered one of the underwater doors to his living quarters.

So the lodge was occupied, the pond was an active beaverwork! I had received my first penny's worth of action—one fat beaver inspecting a dam—and I was happy. Already I was spinning theories about who this individual might be. Judging by his size, he had to be mature, of breeding age, and so perhaps had parented young. Were there kits in the lodge? Had I found a colony?

I could not have succeeded in maintaining even a semblance of calm had I known then that the creature I had just sighted would prove to be the founding patriarch of a dynasty of beavers, animals whose lives I would closely moniter in all weather, during every season, over a period of four years. At the moment, it was all I could handle just to know that Lily Pond was going to be a wonderful place in which to work.

Chapter Two

Every evening the big beaver performed the same ritual. At precisely 6:10, he exited his lodge through an underwater plunge hole, surfaced, and floated in place for a minute, perhaps to allow his beady eyes to become adjusted to the light. Then he headed directly for the dam, which he thoroughly inspected. Back and forth along its 150-foot length he swam, looking and listening for leaks. When satisfied that no water was spilling over its crest, he dived and traveled underwater along its base, looking for seepage there. His diligent attention to the condition of this amazing structure, which measured five feet high on the downstream side, prompted me to name him the Inspector General.

There is a mysterious quality about the beaver, one that conjures up racial memories of trolls and gnomes and "little people." It isn't just that the animal works magic and can dramatically transform its surroundings during the dark of a single night; its odd and lumpy shape is like that of no other creature. In fact, it looks like some kind of mythical beast put together out of a grab bag of parts belonging to other animals. Its front paws are five-fingered, raccoonlike, and able to manipulate all kinds of material with skill. Its hind feet are totally different from its front ones, big webbed paddles like those of a loon or a duck. Its body is similar to that of a woodchuck who has fattened up in anticipation of a long winter's fast. Its tail might have been taken from a duckbilled platypus; it is flat and paddle-shaped with a surface that is beautifully etched, as if tooled by a skilled leather craftsman.

The North American beaver frequently walks bipedally on its loonlike, webbed hind feet, thus freeing its dextrous, five-fingered forepaws for other uses.

To make sense out of this creature was going to be a much tougher assignment than had been the task of getting to know the coyote or the wild horse or the bobcat. Those species I could relate to. For one thing, each of them had a domestic counterpart (the dog, the horse, and the cat) with which I was already familiar. But the beaver was like no other animal I had known, domestic or wild. Yet I had reason to believe it was as social a creature as any I had studied, for the beaver forms strong family bonds. Each colony, as a rule, is made up of one breeding pair of adults, who remain together for as long as both live; their kits of the year (numbering from one to six), and all the surviving offspring born to them the previous year, called yearlings. Throughout winter these family members, sometimes as many as fourteen, hole up together in a dark lodge and share food from a common larder. To do so without shedding blood, *Castor canadensis* would have to have evolved a number of complex social strategies, such as the capacity to give and solicit care, a means of expressing appeasement and assertiveness, the ability to communicate by contact sounds and display postures, and, above all, a high threshold for the release of aggressive behavior. What intrigued me was evidence I had seen that the beaver not only possesses this social repertoire, but is quite capable of directing it toward nonbeavers, that it sometimes forms relationships with human beings.

Some years earlier I noted that this very thing had happened between Hope Buyukmihci and the wild beavers that inhabit her Unexpected Wildlife Refuge near Newfield, New Jersey. To say that the animals that resided in the refuge pond recognized her would be understating the case. They came when she called. Just as astonishing were the beavers that resided inside the home of the late Dorothy Richards in New York's Adirondack Mountains. I visited Dorothy when she was eighty years old. At the time four adult beavers (one a twenty-four-year-old, weighing sixty pounds) were enjoying the run of the house. Dorothy had converted a downstairs bedroom into a swimming tank, complete with a swinging door that allowed the animals free access to her living room. I watched late one afternoon as, one by one, these residents roused themselves for the night, pushed open the flap door and took over the downstairs. One headed for the fireplace, where he gathered up all the firewood. Holding a stack of kindling between his chin and forelegs, he toddled on his hind legs back to the tank room, and there added it, a stick at a time, to a lodge that was in progress. Had he worn a pair of overalls, he would have looked like a storybook character drawn by Beatrix Potter.

Another member of this strange colony, as unmindful of its wetness as an affectionate spaniel, made his way up Dorothy's outstretched legs and seated himself on her lap, which she had hastily covered with an always-at-the-ready rug. His wheedling sounds told her he wanted an apple from a bowl by her elbow, and this he gently accepted with one of his five-fingered front "hands." After finishing it off, he accepted an ear of corn from Dorothy, and with the dexterity of a human child, lifted it to his mouth and, using both hands to twirl it, consumed the rows of kernels, human style. I remember thinking at the time that the species would be a good candidate for domestication. Likely it was spared that fate because of its physiological need to live in a body of water, the level of which it insists on controlling.

The late Dorothy Richards in 1975 with two of her four house-beavers.

Now it was my hope that the Inspector General would grant me the same privileged relationship that Hope and Dorothy had enjoyed with their beaver friends. At times I imagined he was actually contemplating such a move, for on occasion he would stop swimming and drift, eyes fixed on my motionless form. What was going on in his beaver mind?

But then he would revert to his wild ways and become agitated by my presence. With head elevated and nostrils working, he would paddle rapidly back and forth, testing the air for my scent. Following this display of suspicion, he would stop and focus his membrane-ringed eyes on me, while I sat perfectly still and awaited his verdict. If the wind favored me, he might then lose interest and swim away; for, unlike human beings, beavers do not believe everything they see. On the other hand, if he picked up a whiff of my odor or if I carelessly made a noise, he would draw up his tail until it curled over his spine and then bring it down on the pond's surface with such force that his hind feet recoiled right out of the water and a geyser of spray shot high into the air.

The sound of a tail slap has been likened to the crack of a pistol, but I don't hear it like that. To my ear it has a more hollow sound and arouses in me the physical recollection of all the painful belly flops I suffered when, at age eleven, I tried to teach myself to dive. It is a startling noise, especially when heard within the peaceful context of a beaver pond. A tail slap always takes me by surprise, even at those times when my whole attention is fixed on the animal about to execute it.

What is the beaver trying to say when he slaps the water? Almost everyone who has studied or written on the subject believes it is the animal's way of signaling fellow beavers to take cover, like the urgent alarm cry of a bluejay broadcasting news that a cat is about, or the white flash of a deer's raised tail, warning its kind to take flight. I accept that this is part of the answer, for I have often watched members of my colony dive underwater or swim rapidly away in response to this percussive and far-reaching sound. But not always. Sometimes they remained totally unresponsive to the warning, to the extent that they would continue to munch lily pads, while a family member slapped and slapped. And at other times they would respond, but in a lackadaisical manner, moving off toward deep water at a leisurely pace. Perhaps it matters who it is that slaps out the message. Kits and yearlings do not produce a really walloping sound with their small tails, and frequently their warnings go unheeded. This is probably just as well since as often as not they "cry wolf." By contrast, the Inspector General's slaps were resounding and produced a real effect on the colony, to say nothing of me.

I suspect the tail slap serves more than a single purpose. Besides communicating danger to the colony, it probably functions to drive away whatever disturbance has incited it. A loud slap in conjunction with a heavy spray of water, although not likely to send a human onlooker packing, is nonetheless a mighty startling event. Many prey species utilize distracting noises and sudden moves to create momentary confusion in a predator that is eyeing them, and the beaver, being a prey species, would be no exception. Moreover, a vigorous tail slap, in addition to its value as a scare tactic, would send a clear signal to any prowling wolf or bear or bobcat that the beaver who produced it is no easy mark, but a fit animal, one capable of putting up an aggressive defense. Predators generally pass up prey that exhibit evidence of robust health and psychological vigor. They must. Animals that must kill to live cannot risk incurring even a minor injury that might incapacitate them. A hunting animal that fails to respect assertive signals given off by potential victims is not likely to live long enough to pass on its incautious nature to many offspring.

Over the years that I watched the Lily Pond beavers, I witnessed a number of incidents that support my belief that the tail slap is used not only as a signal to other beavers to take cover, but, on occasion, as an outright expression of aggression toward intruders. For one thing, I noticed that beavers will sometimes swim *toward* whatever is alarming them before letting fly with their tails. I have even seen beavers go out of their way so they could perform this confrontational display before an audience. For example, the In-spector General once swam one hundred yards to slap his tail before a grebe, a most inoffensive aquatic bird. The little grebe, who was dipping and diving for small fish and otherwise minding his own business, was no stranger to the beavers; he frequented their pond

With a slap of its tail a beaver can displace a startling amount of water and create a resounding sound.

and, except on this occasion, his presence there was entirely tolerated. In this instance, however, the Inspector General headed directly for him on a straight course, and, when within two yards, he wheeled to face his perceived adversary and gave the water a mighty thwack. The startled grebe dove so fast that he appeared to dematerialize in thin air and did not surface again before putting an acre of water between himself and the Inspector General.

On another occasion, I watched the same beaver travel a similar distance to confront a white-tailed deer who was standing belly deep in the murky pond, munching on lily pads. Here again he carefully positioned himself so as to be directly in front of the doe he was about to startle. Then he let fly with a whale of a tail slap that seemed to propel the deer right out of the water and send her thrashing into a thicket of laurel. Yet on many nights I had seen the Inspector General and that same deer feeding peacefully within a few feet of each other. John reported having seen a beaver in Maine act out this ritual before a moose, but with no such satisfying result. The animal did not budge. Apparently, a thousand pounds of moose does not take orders from a sixty-pound beaver.

What was the Inspector General up to seeking out and startling such innocuous creatures? Perhaps he possessed an appetite for tail slapping that needed occasional gratification. Either that, or he was endowed with a sense of humor.

More understandable is the beaver's readiness to startle creatures that approach its young. I once watched the Inspector General respond to the complaints of a kit who was out of sorts over the near presence of another pond regular, a Canada goose. The baby beaver raised a vocal ruckus that brought the Inspector General one hundred and some yards to the rescue. The big beaver then employed his usual tactic and showed the goose what a beaver's tail is good for, in short order, clearing the area of the harmless bird's unwanted presence. The next day that same goose was back and the Inspector General took no particular notice of him.

On still another occasion I observed one of my beavers direct this kind of behavior at an otter. That beaver, a female, seemed to be caught up in a slapping frenzy, and it took me some time to locate the cause of her excitement. Finally, through my binoculars, I spied an otter porpoising about in front of the lodge. At the time there were newborn kits inside, and this must have accounted for her extreme agitation; for on most occasions, my beavers and the otters that frequented Lily Pond seemed content to tolerate one another. It would have to be so. A beaver cannot slap the water perpetually, and a fish-eating otter normally poses no threat to a beaver colony.

I regretted that my own presence continued to alarm the Inspector General and hoped that in time he would come to view me with indifference. Gradually this did begin to happen. As the summer progressed, his uneasiness subsided. No longer did the sight and sound and scent of me cause him to raise his head out of the water and swim in fast circles. Sometimes he would even risk hauling himself out of the water to scrutinize the dam at point blank range or to feed on some tasty herb that grew along the shore.

It is startling to see a beaver of such size emerge from the water. I have no doubt that he weighed sixty pounds. When swimming, most of this hulk rode beneath the surface and so was concealed from view; but when he mounted the bank, his Falstaffian proportions became evident. Small wonder that *Castor canadensis* took

The Inspector General is the first beaver to tolerate my close approach.

to life in the water, which element buoys up a number of obese mammals, such as the whale and the walrus.

On land the Inspector General appeared ungainly. When sitting up, he tucked his flat tail under him so that it protruded forward between his webbed hind feet. When resting on all fours, his rounded back mounded so high it seemed to dwarf even his huge head. While moving about, his leathery tail, having grown to an impressive length, seemed a weighty drag, like some kind of garden tool used to level the ground. Even so, when surprised, he was capable of executing a most energetic sideways hop, a movement peculiar to beavers. Then he would dash for water at a surprising clip.

Eventually, after I had become familiar with other members of the colony, I came to realize what a fine-looking specimen the Inspector General was. His coat was an even mahogany color, and he kept it groomed to perfection. Though his large size clearly indicated that he had lived for a substantial number of years, this fact was not evident from his appearance. His muzzle bore no trace of gray, nor were there any battle scars on his beautifully etched tail. And when he sat up on his haunches and let his jaw go slack, thus revealing his long and bright-orange incisors, I saw that they were in excellent shape.

The Inspector General was the first beaver at Lily Pond to tolerate me within viewing distance, and for this reason alone I will always feel a special affection for him. Still, even he was in no hurry to let me in on the secrets of his private life. I watched him swim along the channels the colony had made. I observed him feeding on lily pads and grazing the bank. And, of course, I noted his regular inspection tours of the dam. But how he interacted with other beavers, what special role he played in the colony, if any, remained a mystery. Getting to know this odd animal and his companions was going to take a long, long time. Meanwhile, I was grateful to him for what help he was giving me—even if he did dispense it a penny's worth at a time.

Chapter Three

E arly on, I began referring to the Inspector General as "he," but I had no way of knowing this animal's sex. A beaver's sex organs are concealed in an opening in its abdomen, called the cloaca. This pocket also contains other items of importance to the species. Two pear-shaped glands, which produce a kind of all-purpose substance called castoreum, are located here, as are the animal's anal glands. Substances produced by these two pairs of glands are perhaps used in grooming and are definitely used to post information at various places inside the animal's territory. Usually, the opening to the cloaca, which is encircled by a sphincter muscle, is kept closed; thus the beaver keeps his or her sexual identity a secret—except, of course, from other beavers.

Sexing beavers can be done if a researcher is willing and able to perform exploratory surgery. That excluded me. I tried to come up with other ideas. I was aware that Hope Buyukmihci and Dorothy Richards, both of whom had kept pet beavers, had on rare occasions caught glimpses of a penis extruding from a cloaca. But I could not count on making such a serendipitous observation of a wild beaver. Palpating for the penis bone, a technique that requires tactile knowledge of the complicated contents of the animal's cloaca, is a method that has been tried by some researchers desperate to know the sexual identity of their subjects. This idea appealed to me about as much as performing surgery. A more reliable procedure, one less intrusive than surgery or palpation, was discovered by biologists Joseph Larson and Stephen Knapp. By examining blood smears, these researchers determined that a percentage of the polymorphonuclear neutrophil leucocytes in the blood of the female beaver exhibit a peculiar drumstick-shaped appendage. Only an occasional such appendage is seen in blood samples taken from male beavers. But even this method, were I to employ it, would require that I live-trap animals to obtain blood smears, an act not likely to endear me to the beavers whose trust I was hoping to win. For the time being at least, I decided to rely on my intuition, and so I began referring to the sole beaver that would come anywhere near me as "he." At some later date, I told myself, should I notice telling sexual behavior on "his" part, or perhaps enlarged mammary glands on "her" abdomen, I could straighten out my notes to agree with that revelation.

Meanwhile, I was experiencing other difficulties. A month had passed since John had returned to Massachusetts and I had begun my watch in earnest. Every night I had remained at the pond until midnight or one o'clock in the morning, during which time I ascertained that three beavers, in addition to the Inspector General, lived there. To discover this, I had to keep all four animals in sight simultaneously (a feat easier to tell about than to do), for I quickly realized that any beaver that popped into view at the far end of the pond might or might not be the same animal I had counted seconds earlier at the near end. Although beavers move at a leisurely pace when swimming on the surface, once underwater they scoot along like fish. Moreover, the various members of the colony were constantly entering and exiting the lodge. Since I could not yet tell any of them apart, I had no way of knowing whether an emerging beaver had made a previous debut and become part of an earlier count or was an animal who had just awakened, one I had not yet included in any tally. Therefore I had to assume that the Lily Pond beaver population was no higher than the most beavers I could keep in sight at one time. And that number was four.

Learning to distinguish one beaver from another was yet another hurdle. Although the Inspector General was considerably larger than the others, even that fact was not evident at any distance. Unless two beavers were juxtaposed in the water, there simply was no yardstick by which I could judge their relative sizes. Separately, each appeared the same, due to a trick of the brain: Larger-than-average beavers are perceived as being closer than they actually are, while smaller-than-average beavers appear farther away.

The use of binoculars actually compounds this difficulty. Every beaver, when viewed through optics, looks huge, even tiny kits. To determine the real size of an isolated animal, one must view it with the naked eye and fairly close-up, within a few yards. And in the beginning, only the Inspector General would approach near enough to amaze me with his voluminous dimensions. The other three hugged whatever shore was farthest from wherever I happened to be. At such distances, they looked enough alike to be clones.

Several people, hearing of my troubles, suggested I radio collar the animals. But radio telemetry, although it would certainly enable me to distinguish one animal from another, would not solve my basic problem. I needed to get close enough to my subjects to observe their behavior and interactions, and radio telemetry does not help one do this. On the contrary, what radio telemetry allows one to do is track the movements of *unseen* animals from a remote place. Were I to carry out a telemetry study lasting several years, I might see my beavers only once and then in a live-trap. For after being fitted with radio collars and having their ears tagged, the animals would be set free, whereas my own movements would henceforth be restricted to those locations where I could best receive the radio frequencies their collars transmitted.

Of course, from this information I could map the animals' movements and thus discover their travel routes, pinpoint where and when beaver encounters occurred, note the disappearance of any animal from the territory, and gain other such information. This effort, however, would not produce answers to the questions that most interested me. For example, were I to receive simultaneous radio signals from two of my collared beavers at a single location, all this could really tell me would be that the animals had met. What happened during their encounter would remain an unknown. Did they hiss? Did they play? Did they groom one another? Did they cut a tree? Or did they simply pass without so much as a grunt or a sniff? To know what takes place requires someone be present and close enough to see the animals well. My past studies, particularly the years I had spent tracking bobcats, taught me that

Beavers are hard to spot in the dense mat of lilies that covers the pond.

one must earn this kind of on-scene viewing with time spent. What I had need of then was not technology, but patience.

Still, even in the early weeks of my study when the beavers were keeping their distance, I was able to gather some information about the four animals I had counted. Clearly they were lily eaters. From the time they exited their lodge, between six and seven in the evening, until the long summer twilight faded to night, they fed on the buds and pads and flowers of *Nymphea odorata*. Loud munching sounds often were the single clue that helped me locate a beaver I might otherwise have missed.

I now saw how misguided John and I had been in worrying that this colony was in imminent danger of running out of food. The beavers at this pond obviously did not miss the kinds of trees we had been so desirous of finding. What we had failed to appreciate was the plasticity of *Castor canadensis*, which survives in climes as disparate as Nevada's sagebrush desert, Louisiana's swamps, and Alaska's taiga. Our perception of the animal's needs had been based on literature and on John's long years of beaver-watching in New England. The species, as it turned out, was far more pliant than

A beaver rolls a lily pad into double scrolls for easier eating.

either of us had suspected. Still, I wondered how the four beavers would fare during the winter months, after ice covered the pond and locked them into their lodge for three months. Lily pads, unlike aspen branches, do not keep in storage.

Meanwhile I was amazed by how this single food item dominated the colony's diet. For at least an hour after emerging from their lodge, they did little else but gorge on this ubiquitous plant. And even as the evening wore on and they became involved in other activities, they frequently snacked on it, sometimes consuming a floppy leaf while on the move. Their ability to do so was impressive. While swimming along channels of open water, they would reach out with a satiny black hand to pluck a single one from the solid tangle of vegetation that lined their route and, without losing a stroke, continue on. While paddling with their hind feet, they used their front paws to roll the big rubbery pad into a double scroll, then slowly inserted it into their chomping mouths.

Unlike many wild animals, beavers do not bolt their food. On the contrary, they hold their leaves and flowers and sticks with dainty fingers, and chew each bite thoroughly. When a number of

Beavers rely on their noses to identify vegetation.

beavers feed together on a tree they have felled, they look and act like a party of well-bred gnomes enjoying a picnic. Their highly developed tactile sense contributes to this illusion, for to a surprising extent, each animal keeps its own place during the feast and feels about with its hands for extra helpings that are within reach. A beaver's use of this keen faculty is not unlike that of a blind person. Beavers, in fact, sometimes behave as if they were sightless. For example, members of my colony often did not so much as glance at the lily pads they harvested, but instead felt about underwater with their sensitive hands for a stem to pluck. Then with heads tilted skyward and side-set eyes focused on nothing at all, they adeptly rolled and inserted into their mouths the single leaf they had picked. And even on those occasions when their tongues announced an unexpected taste experience, they still did not take a look at what they had in hand, but lifted the leaf or stick or flower to their noses for a quick scent check.

I don't mean to suggest that a beaver does not use its eyes, which are well adapted for life in water. Each is ringed with a transparent membrane that slides across it whenever the animal dives. Thus equipped with goggles, the animal's eyeballs are protected from debris and muck, while at the same time light is able to penetrate this nictitating lid and fall onto the animal's retinae. On land too the beaver makes good use of vision. I have seen beavers study the tops of trees they were about to topple, lock eyes with one another, scrutinize their dam crest for breaks and study me. Still, I cannot shake the impression that they do not entirely believe what they see; for in many situations they employ touch or smell or hearing when vision would seem to do as well.

Swiss anatomist George Pilleri, who made a detailed study of the European beaver's brain, found the animal's optic nerve to be quite narrow in relation to eye size, which, at that, is small. By contrast, he found its olfactory faculty to be unusually developed, having suffered no regression as a result of the species' adaptation to aquatic life. This is contrary to what has happened to many other land mammals that returned to life in water, namely whales and seals. Pilleri went so far as to suggest that beavers may even make use of their acute sense of smell while swimming *underwater*!

In thinking about all this, it suddenly struck me that a beaver would have little need of vision inside a dark lodge, where light does not penetrate, and so might lose the habit of relying on its eyes. Inside a dark lodge, of course, is where a family of beavers spends nearly all the daylight hours. Moreover, throughout winter the colony is sealed by snow and ice into a subaqueous world. Since the species does not hibernate, wintering beavers must somehow manage to move about in their black-as-night dwelling and feed and groom and so on without getting in one another's way. No wonder when the animal emerges into the light it goes right on groping for objects it could as easily locate by sight. No wonder it feels about for lily leaves as if playing a game of blindman's buff. And no wonder the plain sight of me sitting beside the pond often aroused only moderate alarm in the Inspector General—that is, until a gust of wind or some noise I made confirmed to him the reality of what his eyes were registering.

The fact that the beaver appears to rely so much on its tactile, olfactory, and acoustical senses is not unlike the trade-off we human beings made in electing to depend on sight and hearing over our senses of touch and smell. We too have been shaped by past and present conditions. Our tree-dwelling ancestors had to trust their vision when making leaps from branch to branch at dizzying heights. And later, after opting for life on the ground, being abroad in bright daylight must have seemed a lot safer than joining the ranks of predators that hunt by night. As a result, we eye everything, believe what we see, consume words, are impressed by appearances, entertain ourselves with flickering images. Had our ancestors spent most of their waking lives in dark or dim settings, things might have been otherwise. In that event, I might now be writing about *Castor canadensis* with smears of lily pollen, blobs of musky beaver-scent, and the fertile odor of pond muck. My dogs would have liked that. The language of scent, after all, is more universally understood than that of the coded word, even though *Homo sapiens* hardly comprehends it.

Chapter Four

D uring summer, daylight lingered on past nine o'clock, giving me three hours of illumination by which to view my colony. Still, I remained at the pond, sometimes until one or two in the morning. Sitting in the dark every night not only taught me to make better use of my ears, it also acquainted me with another plane of existence. But first I had to make friends with night. First I had to lay to rest the ghosts that haunt the darkened rooms of children, those racial memories of chimerical beasts, which I early learned could be magically exorcised with the flick of a light switch. First I had to overcome my fear of being alone in the park in the dark. At the same time, I had to allow my occasionally overactive imagination enough freedom to concoct visual scenarios to fit the strange night noises I was hearing. What animal could be making such a piercing cry? What might be causing that dragging sound? Why did all the insects suddenly stop humming? Was that a football behind me?

My ability to see in the dark varied from one night to the next, depending on the phase of the moon, the amount of cloud cover, and perhaps the number of carrots I had eaten that week. When the moon was big and bright, I could see well enough to confirm some shot-in-the-dark assumptions I had made. For example, I established that the tiny creature that occasionally scurried across my boots was indeed a mouse. But I was in for some surprises too. Against a lunar sky, I made out the silhouette of a recently fledged great horned owl, thus resolving the "mystery of the guttural shrieks"—sounds that had for several nights conjured up images in my mind of a person being strangled. Who would have imagined that a young hooter could make such a squack? Yet I saw that the young bird's hideous cries served it well, for they brought a food-bearing parent to the dead tree snag upon which it perched. On a moonlit night I was also able to discover the source of another eerie racket, raccoons squabbling over crayfish.

The beavers were as silent as they were invisible. Sometimes I could hear one slip quietly into the water, and only then did I realize that he had been foraging along the bank quite near to where I was sitting. Unlike myself, they did not splash about while swimming. Even when diving, an action that required them to counter their natural buoyancy with a hard thrust of their webbed hind feet, they did so silently—expending no unnecessary energy, wasting no motion, so that the entire operation was no more audible than the dip of a canoe paddle wielded by an Indian.

Their quietness was all the more amazing, given my suspicion that they were no better equipped to see in the dark than I. This idea hit me early on, while making preliminary observations of beavers in the New Jersey Pine Barrens and in Massachusetts. In trying to photograph some animals there, I made use of strobe lights, which flash only briefly and in synchrony with the camera's shutter-release button. As the evening light faded, however, I found it necessary to switch on an ordinary light and shine it on my subjects for as long as it took me to focus my lens on them. At the time I was laboring under a common misconception that nocturnal animals do not see red light, so I covered the head of this focusing light with colored celluloid.

Despite my error in this, the system seemed to work. The beavers did not appear to notice the red light. Nor were they bothered by the fast firing of my strobe, which was activated for less duration than a bolt of lightning. One night, however, the celluloid fell off my focusing light, and the beavers were bathed in its white beam. To my surprise, they did not seem at all disturbed by this.

I speculated that the animals had probably become habituated to the headlights of passing automobiles and, from then on, I left my focusing light uncovered. As a result, I discovered something else: when caught in a beam of light, a beaver's eyes do not look

Though lacking night vision, the beaver functions well in the dark.

like two burning coals. This was significant information, and there could be only one explanation for it: the beaver's retinae do not contain the light-gathering crystal called *tapetum lucidum*.

Why had the beaver not evolved this adaptation, which is present in raccoons and wolves and bobcats and deer and possum and most other nocturnal mammals? Could it be that the species is too new to life at night to have acquired it? I researched records of early trappers and found evidence to support this hypothesis. Several early observers of the animal described it as being diurnal, even reported that it liked to sun itself atop its lodge. *Sun itself?*

If these descriptions can be believed, they raise another question: what would cause a diurnal species to become a nocturnal one? Could such a change have come about as a result of the extraordinary trapping pressure exerted on the beaver over three centuries of North American history?

No animal has suffered such relentless exploitation as has the beaver. Given the history of North American exploration and settlement, it is something of a miracle that the species is still extant. In the early seventeenth century the French and the Dutch were

no less single-minded in their quest for riches than were the Span-
iards, who at the time were plundering Mexico and South America
of gold. The French and the Dutch, however, focused their attention on furbearing animals and concentrated their activity on the
waterways, where they discovered beaver in vast numbers. Beaver
products, both fur and castoreum (then believed a cure for everything from frostbite to hysteria) were much in demand in Europe,
where no well-dressed gentleman was considered properly attired
without a beaver top hat. To satisfy this craze, beavers had already
been hunted to near extinction on the European continent, and the
Dutch settlers in America, ever alert to trade opportunities, were
quick to respond to the ongoing demand from abroad for beaver
hats. In short order, Fort Orange on the Hudson River (today's
Albany) became a leading port for shipping furs and was subse-
quently rechristened Beverwyk. As for the pelts themselves, they
became a kind of gold standard of the period and were accepted
everywhere in lieu of coin.

Meanwhile, to the north, the Montreal-based French were also
seized with beaver fever, and they began sending their *bateaux*
southward into the streams and lakes of the Adirondacks in search
of the animal. Thus began a bitter struggle for control of the region.
Both sides enlisted the help of Indians, not only as guides but as
suppliers of beaver pelts. While some Native Americans may have
resisted killing an animal traditionally associated with the creation
of the world, a number of tribes had made prior use of the beaver
for food and ornament, and were skilled at hunting it. And so, in
exchange for European goods, they joined in the plunder of the
region.

When Samuel de Champlain first visited the Adirondacks, "every
lake and pond was occupied—every river, brook and rill, from the
largest to the most insignificant, was thickly peopled" with *Castor
canadensis.* Some half century later, in 1672, an estimated one million
beavers probably still existed in that rich 16,000-square-mile mountain range.* Once the downturn in beaver numbers began, how-

---

*The estimates on historical beaver numbers given in this chapter were published
in 1908 by New York's Forest, Fish, and Game Commissioner and were based
in part on extrapolations from annual pelt exports.

ever, it would not be reversed, for no restraints were placed on the beaver hunters, and so excessive was their take that, on one occasion, 75 percent of the pelts they brought to a glutted auction were burned to hold the price of the fur at a profitable level. As a result, by the year 1800, the Adirondack beaver population had been reduced by more than 99 percent, to only about five thousand animals.

Long before that year, of course, the fur hunters had extended their activities to distant parts of an expanding nation. And everywhere they went, their passage was marked by drained ponds and collapsing beaver lodges. To destroy entire colonies was relatively easy, inasmuch as each beaver family inhabited a highly visible dwelling. From earliest times these sturdy fortresses, many built in the middle of ponds, had protected the species from its natural enemies. With the arrival of Europeans in America, however, the animal's strategy backfired. A large and conspicuous shelter did not enhance survival; on the contrary, it invited annihilation.

Over the next twenty years, beaver numbers in the Adirondacks fell by 80 percent to a mere one thousand animals. Elsewhere the situation was similar. In 1811, John Jacob Astor, dissatisfied with a declining pelt take in the Great Lakes region, sent an expedition to Oregon Territory to search for beavers in that rugged clime. But his famed Astoria fur-trading post at the mouth of the Columbia River was a short-lived enterprise. So efficiently did he plan for the capture of every last beaver along every last tributary and every last stream feeding into the Columbia River that inevitably the golden age of the fur trade collapsed for want of the very creature necessary to sustain it.

Then in the 1840s the steel-jawed leghold trap was invented, a device that made it possible for free-lance trappers to get into the act and wring a last bit of profit from the nation's severely depleted beaver population. As a result, before the century ended, the species was such a rarity, so near extinction, that most people proclaimed it gone. In 1895 only five beavers were known to exist in all of New York state, and their whereabouts was kept a secret. Fourteen states—Massachussets, Vermont, New Hampshire, Rhode Island, Connecticut, Pennsylvania, New Jersey, Delaware, Maryland, Illinois, Indiana, Ohio, West Virginia, and Florida—announced they had no beavers at all. The remaining states had poor or no information on what remnant populations may have persisted within their borders.

One can speculate that what few animals escaped this continent-wide decimation must have been the wariest of their kind, deviants, disinclined to build conspicuous lodges. And indeed late nine-teenth-century reports of sightings describe the beaver as a recluse, an exclusive bank dweller. One can also speculate that these sur-vivors may have escaped the notice of trappers by turning night into day, for by the end of the last century, no further mention is made of beavers sunning themselves on their lodges.*

The phenomenal comeback of the beaver across North America is a success story conservationists can point to with pride. It rep-resents one of the first efforts by Americans to save an endangered species from extinction. Such an undertaking would have failed had it not been for a small group of people who preached a new attitude toward nature and who lobbied their state legislatures for laws to protect the beaver. They also devised and carried out a plan to promote the animal's recovery. First they searched for and captured what beavers could be found. These were planted in suitable and protected habitats. (Seven captive beavers, on loan from Canada for exhibition at the Louisiana Purchase Exposition at St. Louis in 1904, were afterward used for this purpose.) To the delight of all involved, these recovery efforts proved effective, particularly in New York's Adirondack Mountains, where the animal immediately bred and began spreading through the region. Today the beaver is relatively common and is once again found in every state of the union.

And so it seems logical to ask: did those few surviving animals who were the progenitors of all present-day beavers possess a quirky biorhythm? Were their activity cycles exactly reversed? If so, was this deviant biorhythm passed on genetically to their offspring, thus causing a permanent change in the behavior of the species? It might be so, for that is how natural selection works.

I studied the eyes of other nocturnal creatures. The owl has incredible night vision, despite the fact that its retinae, like the

---

*This behavior has recently been reported to me, having been observed once again in a protected population in the Adirondacks, of all places!

beaver's, contain no light-gathering crystals. Yet its eyes are elegantly designed to see in the dark, for they are huge in proportion to the bird's head; moreover, in dim conditions their pupils dilate and all but fill the bird's eye sockets. By contrast, the beaver's eyes are tiny in proportion to its head.

Indeed, by now I was convinced that the beavers I was watching possessed no better night vision than I. Yet, unlike myself, they did not stumble over tree stumps or fall into holes while moving about in the dark. I wondered if in time I would develop some special faculty to help me get around so well.

Meanwhile, the world of darkness was growing on me. I even began to look forward to its onset. As twilight faded, I welcomed the chirps and scrapes of insects as they tuned up to play their repetitive night music. I listened for the tremulous whinny of the common, though rarely seen, screech owl. I looked forward to the magic moment when the beaver-created channels on Lily Pond would perfectly mirror the night sky. In short, I came to feel at peace in the dark.

Yet with summer passing and daylight shrinking, I grew increasingly uneasy. I was not getting on with the task of observing beaver life. As a start, I needed some way of telling the four animals apart. To accomplish even this much required I view them close-up, together, and soon. There was no helping it: I would have to use food to lure these elusive subjects to the cove.

Chapter Five

Since all four beavers were growing fat on a superabundance of lilies and had no need to risk coming ashore to satisfy their hunger, I tempted them with aspen, a beaver treat that did not grow at Lily Pond.

I don't believe in feeding wildlife, but like most people I don't always abide by my own principles. In winter I regularly put out seeds for cardinals and chickadees and nuthatches and downy woodpeckers, and delight in watching them dine on my handout. Does my tampering with the natural limiting effects of winter on bird populations do more harm than good? Some say yes. Purists in this regard believe in letting winter take its toll of the surplus birds that have hatched earlier in the year; they argue that otherwise, come spring, many birds will not find suitable habitat or nest sites and, as a result, will die anyway from other causes.

Other people make the opposite case. Winter may be nature's way of limiting animal numbers, they say, but bird populations today are not being regulated by nature; they are being decimated by unnatural, man-related activities, such as power lines, chemically treated grains, polluted streams, high-speed traffic, and insect spraying. No need to worry that too many birds will survive due to man's helping hand; the real problem is that too many birds are dying as a result of man's unnatural impact on natural systems. A little feeding during winter cannot begin to offset the negative consequences our way of life imposes on wild birds.

This argument might as well be extended to beavers. Man makes life difficult for *Castor canadensis* when he rips out its dams to prevent roads and fields from becoming flooded or when he wraps screens around trees to protect them from being felled and eaten. I reasoned that what few branches I occasionally donated to my colony could not outweigh the pervasive and negative impact of man on beavers. In any case, the fate of my colony did not hang on what little I brought them. The animals were gorging on a bumper crop of lilies. Should they discover my gift of aspen, they might eat that too, just as I experience no difficulty in making room for a tasty dessert.

To obtain the aspen I needed, I enlisted the help of a friend, Dan Pierson, on whose property the tree grew in abundance. Dan was interested in seeing the animals I was studying, and so I invited him to come with me to the pond. The two of us concealed ourselves behind a huge boulder just a few feet from where I placed the branches in the water. While we waited, he told of watching beavers as a young boy. Once he had observed two beavers climb out of the water and perch on a dam, where, side by side, they nuzzled one another's face and made soft whimpering sounds.

"They looked like they were kissing," he said.

I was not surprised by his account. Such displays of affection have been reported by beaver-watchers as divergent in their philosophies as Dorothy Richards, who kept the animals as pets, and Lars Wilsson, a Swedish biologist and strict determinist, whose study of the European beaver has been cited by many as the definitive work on the subject. Wilsson described the behavior of a courting pair as follows:

"They sleep curled up close together during the daytime and at night they seek each other out at regular intervals to groom one another or just simply to sit close side by side and 'talk' for a little while in special contact sounds, the tones and nuances of which seem to a human expressive of nothing but intimacy and affection."

It was just this kind of social interaction I most wanted to observe, but thus far my subjects had not let me see anything so charming. The most intimate behavior I had witnessed was what I called the "Eskimo kiss," when two members of my colony would swim toward each other, touch noses for a few seconds, then break apart.

My view of the animals was of course limited to what took place outside their lodge. How they behaved inside their cramped quarters was undoubtedly more revealing of their nature. For a species that lives in such close contact must have evolved social strategies to prevent it from commiting matricide, patricide, and infanticide. The beaver's sociable nature had to be a fact.

This conviction was based partly on the fact that *Castor canadensis* has diverged so radically from the overwhelming majority of species that fall within the order *Rodentia*. Unlike the mouse, for example, whose family bonds last all of eleven days, after which time it casts its fast-maturing young out of the nest and gives birth to more, the beaver invests much time and energy in the raising of its single annual litter of offspring. Beaver kits remain at their natal pond for two years and sometimes longer. These very different lifestyles represent contrasting strategies by which the mouse and the beaver attain the same end—the survival of their species. But whereas the prolific mouse bets that by producing a great many progeny some will survive, the slow-breeding beaver operates on the opposite premise. It counts on a small number of well-prepared offspring to perpetuate the family DNA. Normally, a species that enjoys a long dependency on its elders is highly intelligent and possesses an interesting repertoire of social strategies as well. Thus I expected to see such social characteristics in the animals I was watching. But my night-loving subjects were cryptic creatures. They performed their miracles and carried on many of their transactions underwater, under cover of darkness, inside their lodge, or just plain beyond the limits of human vision. And now I, like some Scandinavian peasant who places a bowl of milk in the barnyard to propitiate unseen trolls who do mischief by night, was now hiding and waiting for beavers to approach my offering of aspen.

Dan seemed amenable to staying at the pond long after the sun had set, when about all we could see was the play of the moon on channels of open water. Nevertheless, our senses were heightened in the dark. Every burble and ripple registered acoustical information in our brains, and we became keenly aware of the pungent pond smells as we sat in silence, communing with our surroundings.

Then we heard it—the sound of gnawing, a staccato scraping of teeth against wood, and no more distant from where we were hiding than either of us could jump. We peered hard into the blackness and strained our eyes. Gradually, by acquiring the trick of looking

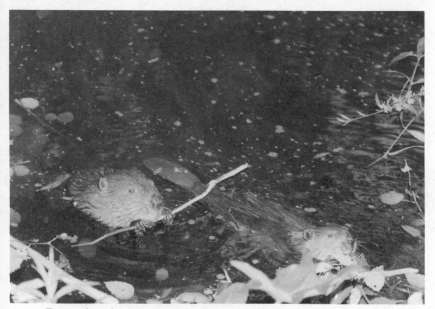
Two age-classes show up at my viewing station.

somewhat obliquely at the animals I wished to see, I was able to make out four forms. Two were large; two were medium sized. Now that the four beavers had lined up in front of me, I was able to ascertain that the colony consisted of a pair of breeding adults and their yearling offspring, apparently born in the spring of the previous year.

We remained at the pond far into the night while these animals came and went and then returned again to clip sticks from the tangled heap of branches that we had set out for them. Often they fed side by side with only an occasional vocal protest uttered. This increased in volume when one beaver, while seeking a fresh helping from the food pile, passed too close to another beaver, causing him to become possessive of the stick he was debarking. If such a vocal warning was not immediately and properly respected, the feeder would make a quick lunge, a mere feint, toward the offender, who would then back off in a hurry. Though a beaver is capable of hissing and growling and does so when the situation calls for it, not once did I hear that level of aggressivity expressed. On the contrary, the beavers' vocalizations sounded more like entreaties than threats and called to mind the utterings of infants without language, who are nonetheless able to effectively intone their wishes.

"Uh, uh, uh, uh!" Translated, the animals seemed to be saying: "My stick, my stick, my stick!"

It was an eventful night despite the difficulty I had making out the scene, and I felt greatly encouraged. I had at last been presented with a tableau of beaver life. I had observed a family of beavers, feasting on aspen—sharing, communicating, and taking care to avoid actual confrontation. Like the many coyote packs I had watched feed together on carcasses, each beaver was protective of his or her space while at the same time cautious not to intrude on the space of others.

However unlike are those two species, the coyote and the beaver share three things in common: both possess lethal weapons—sharp teeth, capable of doing murderous damage; both possess strong inhibitions against using those weapons on their kin; and both possess social strategies to prevent those inhibitions from being over-ridden. While the coyote ritualizes his aggressivity by opening his mouth in a wide gape, the beaver expresses his distress vocally or makes mock lunges at any relative who ignores his warning.

It may have been only one more penny's worth of observations, but to this animal-watcher the night's revelations counted as a big and long-overdue reward.

Chapter Six

To a beaver-watcher, the autumnal equinox, which occurs on September 22 or 23 and marks the end of summer, has real significance. On that date, day and night are of equal length. From then on, darkness progressively outstrips daylight with every rotation of the earth until in late December the situation begins to reverse itself.

During the last days of September, the beavers' sleep-wake schedule became erratic. On some nights animals would pop out of the lodge as early as five o'clock, giving me more than an hour and a half of illuminated viewing time. On other nights, they did not emerge until long past seven, after the sun had set. To keep tabs on them, I had to make use of artificial light, a high-beamed halogen lantern, which I directed at any sound that might be caused by beaver activity. On such nights I tried to lure the animals into close range with token-offerings of aspen. For even in the dark, they were always visible at short distances.

By now the animals had become moderately indifferent to my presence at the pond. Sometimes when I stood on their dam they glided back and forth at my feet. At other times I was allowed to trail behind a beaver traveling the donut ring (the name I gave the beaver-made channel that encircled the pond). And on a few occasions, one or another would even leave the safety of water and climb up on land to graze not far from where I was standing or seated. Oddly, that animal would be less startled at discovering my near presence than when I was some distance off. Were they by

Curious beavers make their own study of me.

now able to recognize me at close range? And were they by now reassured that I would do them no harm? Or had they, as a result of my ubiquitous presence, become dangerously trusting of human beings in general? Could a wild beaver distinguish one person from another?

These questions troubled me as the animals began to take an active interest in me. One night as I was sitting peacefully in the dark, listening to the lap of water against the rocky shore, a large dark form hauled out of the pond and waddled up the bank to within three feet of me. At such close range, I could see it was the adult beaver I had named Lily—an animal I knew to be the Inspector General's mate. I had, in fact, recently caught sight of those two sitting on a rock, carrying on quite like the pair Dan had told me about. For several minutes, they had rubbed their faces together and "talked."

Now Lily cocked her head at a three-quarter angle and viewed me out of one eye. I spoke to her in a reassuring tone and took care to hold absolutely still. It was a bit disconcerting being so close to an animal whose teeth could cut through a mature hardwood tree in a matter of minutes. What was on her mind?

When Lily had seen all she wanted, she turned, waddled down the bank, and slipped quietly into the water. As she disappeared into the dark, my adrenaline stopped pumping, but my mind continued to race. What had been her motive in coming up to me like that? Being studied by a beaver gave me a taste of my own medicine, and I had a new appreciation of how perplexing and alarming it must be to an animal to be so closely watched by me. The experience raised doubts about what I was up to, and I didn't know whether to feel elation or remorse over what had just happened.

For some nights thereafter, I brought no branches to the pond and tried to remain downwind and out of sight and hearing range of my beavers. This approach, however, brought me absolutely nothing in the way of observations; thus my whole purpose for being at the pond was defeated. So once again I carried a small bundle of aspen branches and placed it in the cove below my perching rock. On that particular night I invited Dan Pierson to accompany me, and he took a position a bit uphill from mine. While waiting for the beavers to discover this offering, I let Dan know the names I had given each animal. The two adults I called the Inspector General and Lily. Their yearling offspring I named Laurel and the Skipper.

Laurel and the Skipper, I explained, often traveled together, or at least within sensing distance of one another. On occasion they porpoised about, diving and surfacing in amazing unison—sometimes in tandem, sometimes one behind the other. At other times, they sped toward one another on a collision course and then, just as it seemed certain they would bump, both dived at the same instant.

I could not promise Dan that he would see such things, however. Unless the beavers would come out at once, the most we could expect to take in would be the sounds and vague outlines of four beavers feeding in the dark. By now it was early October and viewing time was brief indeed.

But on this night the moon came up big and orange and lit up the pond. The beaver-made channels, unruffled by wind or breeze, looked like dark glass, here and there etched with glittering Vs— the wakes of swimming animals. Even the beavers' wet coats sparkled in the moonlight.

"It looks like I've sprayed them with some kind of iridescent paint," I commented, then added, "and that might not be such a bad idea either."

Dan nodded and then we sat in silence. Conversation could only mar the night's perfection. How many such spells had I missed during the light-oriented existence I led on earth? How was it that I had become such a slave of the day, so attracted to the light, so bent on electrifying the darkness? It wasn't always so for the human race. The invention of the light bulb, barely one hundred years ago, has made us strangers to night. Neon signs and flood lights now blot out the stars.

A few waves lapped the shore, set in motion by an animal swimming underwater.

"They're here," I whispered over my shoulder to Dan and then leaned forward to see which beaver would emerge.

"Why it's a baby!" I all but said this aloud, so surprised was I.

"A baby? A baby what?"

Dan crawled down to my rock and leaned over the water.

"What do you know, it's a baby beaver!" he said, trying to stifle his chuckles.

"This is impossible. Baby beavers aren't born in the fall. This is October!" I said.

"Well, that's a baby beaver," Dan said in a voice bubbling with suppressed laughter.

"Yes, it is," I agreed. "And there's another one!"

What we were seeing was not supposed to happen. Two infant beavers, perhaps four weeks old, their tails as small as tablespoons and their bodies no bigger than a red squirrel's, were tugging at aspen branches. One clipped off a small piece and then, with its tiny hands, tried to stuff a still attached leaf into its mouth.

A moment later Laurel surfaced and the baby immediately began emitting the same desperate sounds I had heard expressed by the older beavers when trying to discourage others from approaching their sticks: "uh, uh, uuh, uuh, uuH, uUH, UUH, UUUH." ("Stay back! Stay back! My stick! My stick! My stick!")

The baby's anxiety, however, was instantly appeased when Laurel selected a branch for herself from the pile I had left along the shore and settled down beside him to feed. The two sat hunched in the shallow water, holding their sticks with a two-fisted grip

Born out-of-season, this kit makes his debut in October in the company of a yearling.

while they slowly turned and debarked them with their teeth, like a couple of lathe operators.

Meanwhile the second kit had carried a stick a few yards offshore, where he labored to consume it while floating on his belly. To do so, he had to raise his head and forepaws out of the water. Then to counterbalance the front part of his body, he arched his back so that his tiny tail pointed straight up in the air. This odd posture gave him the look of a toy boat, the bow and stern of which had been carved to resemble animal parts.

Kits in the fall? How had this happened?

Despite their immaturity, the little kits demonstrated that their teeth were already in excellent working order. Both clipped twigs, chomped leaves, and stripped bark with surprising facility. Because they did not yet know how to roll leaves into manageable shapes, they simply balled them up and stuffed them into their munching faces. Likely, they had been chewing on fibrous vegetation for some time, for adult beavers bring solid food to lodgebound kits when they are only a few days old. It is something to think about that beaver mothers nurse their offspring for eight or more weeks, even though the babies are born with sharp little incisors already erupted.

59

Kit in the moonlight appears iridescent.

I regretted missing the first weeks of the kits' development, but it could not have been otherwise. During this infantile phase of life they had been kept in the lodge, while every member of the colony catered to their needs. Fresh greens were delivered to them, their bedding of wood chips or grass was regularly changed, and the entire family took turns watching over them, for baby beavers cannot be left alone. Should one fall into a plunge hole and get out into the pond, he or she would not be able to reenter the lodge unaided. Though born with eyes open, fully furred, and able to swim, infant beavers are too buoyant to dive, and since every opening to a beaver lodge lies below the water line, only a diving animal can make use of them.

One important event in the development of the beaver kits—their grand debut into the pond—must have taken place quite recently. When old enough to dive, beaver babies are literally ushered from their snug living quarters into the water. This first excursion is well attended, not only by their parents, but by other members of the colony as well. An entourage of adults and yearlings surrounds the youngsters and provides them handy backs on which to hitch rides in the event that they grow tired.

I judged that the two beaver kits Dan and I were looking at had been out of the lodge for about a week, for they were already able to swim and dive quite well. Nevertheless, they stuck close to Laurel and wailed when she started off without them. Paddling rapidly through the lilies to catch up with her, they looked like two water sprites. At that moment I named one Lotus and the other Blossom.

I hoped those two tender buds would develop rapidly and become hardy enough to make it through the harsh season that soon would be upon them. For that matter, I hoped all the Lily Pond beavers would survive winter. But I was not optimistic. Looking about, I had to wonder what food, if any, was available to sustain the colony during the cold months ahead when lilies no longer would bloom. And now there were six mouths to be fed.

Chapter Seven

By the first of November the kits had not grown much, but they were decidedly more self sufficient. Often they emerged from their mud-and-stick residence unaccompanied by an adult or yearling and spent the first half hour of the evening in the deep open water directly in front of the lodge, cavorting about like a pair of seals. Their antics included much coordinated diving and surfacing and gave me the feeling that I was watching an aquatic ballet, most of which was taking place below the surface. I longed to see their underwater chases, to watch them swim apart and come together, circle and rise—duets I knew were being performed. For after remaining submerged for a time the two would pop up in perfect unison, remain topside just long enough to take a breath of air, and then plunge effortlessly and silently into their watery realm again. When they had had enough of this sport, they swam off to one of their favorite feeding spots, sometimes together, sometimes separately.

I do not mean to suggest that the kits were no longer dependent on their elders. When either spotted an adult or yearling feeding on the pond he or she headed directly for the dining beaver and, after a brief greeting (a ritual that involved nosing the older beaver's side and then sniffing his or her mouth), the youngster would pluck a lily pad and settle down to eat in company. Often the baby positioned himself but a yard away and face-to-face with the more experienced animal. Thus he was able to watch the older beaver eat and likely acquired a certain know-how on how best to handle

The kits pay close attention to the activities of their parents and older siblings.

unwieldy lily pads. Since the young kits did not roll their leaves, sometimes a big floppy pad would "come alive" on a little beaver who was trying to guide it to his mouth; on occasion one would even end up on the baby's head. Watching an adult use both hands simultaneously to convert the flat rubbery pads into manageable double scrolls perhaps helped the kits learn to cope with the problem; for all the young born to my colony sooner or later acquired the knack of double rolling lily leaves.

Undoubtedly, the kits' tendency to tag along after older beavers served a more important function than merely informing them how to manipulate lily pads. In following the more experienced beavers about, the kits were introduced to and acquired a taste for particular foods. Likely every beaver colony exploits a slightly different mix of vegetation. The Lily Pond beavers, for example, had no experience with willow, which constituted the main diet of a colony just one-half mile away. By the same token, those willow-eating neighbors did not eat water lilies, the mainstay of the Lily Pond beavers, for none grew in their pond.

Beavers live in small social groups, or colonies, which occupy

A young beaver feeds on the same branch as his mother.

exclusive areas called home ranges. To some degree, the plant and
animal composition even of contiguous home ranges differs; therefore, the feeding habits of their respective inhabitants are not the
same either. Nor does any colony's diet exactly conform with what
is generally considered to be the standard fare of the species. It is
possible therefore that emigrating offspring, in seeking new areas
to homestead, are influenced in their decisions by the discovery of
foods that are familiar to them. Thus a variety of habitats continue
to be exploited by animals from different "food cultures," maxi-
mizing the species' use of land and minimizing the risk of its becoming addicted to a single preferred food, one that could disappear
during a blight (e.g., the panda's dependence on bamboo).

The fact that beaver kits remain with their family for two years
and sometimes longer certainly suggests that the young have something to learn from their elders. At the same time, it would be hard
to dispute that a good many of the beaver's behavior patterns are
inherited, for the species is a "born builder," and stereotypical
motor sequences used in the construction of dams and lodges are
expressed spontaneously by each individual as he or she matures.

For example, Lotus and Blossom made building motions in thin air long before they helped construct a dam or even so much as looked on while their elders packed mud onto the family lodge. Moreover, beavers raised in captivity and then set free as adults are perfectly able to construct dams and lodges and dredge canals without ever having seen such work done. Nevertheless, I agree with those biologists who maintain that young beavers acquire greater proficiency at these tasks during their unusually long dependency. Like lion cubs, who are also programmed to express certain innate motor sequences—stalking and pouncing, for example—but who nevertheless have difficulty grasping the idea that such behavior must be applied to the practical business of killing food, beaver kits likely need role models to show them where and how best to use their innate building skills. The tendency of young animals to copy their parents has high survival value and there is no question in my mind that the kits I watched imitated their elders.

One night I brought a bundle of aspen to the pond with the intention of luring the beavers into viewing range so I could note more physical differences among the various individuals. I was foiled in this purpose, however, when Lily, instead of remaining at the cove to eat my offering, began towing it, branch by branch, across the pond. Through my binoculars I watched as she systematically went about planting it underwater in front of the lodge. Of course, she was laying in stores for winter, and I was fortunate to have arrived just in time to see her initiate the process.

Because beaver houses are constructed so that all openings lie beneath the water line, a colony must store enough branches on the bottom of a pond to nourish its members over a period of three or four months, for after the pond freezes over, the animals are confined to the interior of their lodge and to what water remains navigable under the ice cover. Quite literally, they are sealed into an aquatic world, unable to go ashore for food.

I was delighted to see Lily's hoarding behavior and I wondered what had finally triggered it. Was it a change in the weather? Or was it in response to the fact that trees were now letting down their sap? Curiously, a few of the branches I had brought, when dropped in the water, had sunk, a phenomenon I had not seen before. What was going on? Without sap in them, the branches ought to have become lighter, not heavier. How was it possible that some of these light pieces of wood now defied physical law? Might this be due to the fact that the pond was on the verge of freezing? When cooled by air to 39 degrees Fahrenheit, surface water reaches its maximum density and sinks. Could this movement be sufficiently forceful to take green wood with it?

The twigginess of the limbs I had brought made it difficult for Lily to maneuver them through the narrow open channel that led to the far side of the pond, even though she gripped each one properly by the butt end and allowed it to drift alongside her body as she swam. When a protruding fork snagged on the mat of lilies that lined both sides of her route, she freed it with a hard yank and then continued on her way.

I shined my lantern on her as she crossed and recrossed the pond, and noted how she pulled each load underwater at a spot she had apparently decided was the right place to plant a winter food cache. Most of the wood she brought could be scuttled with ease, but some effort was required to prevent large pieces from bobbing to the surface, requiring her to remain underwater and out of sight for some time. I tried to picture what was going on down there. Each branch would be driven into the bottom muck, pointy end first and with sufficient force to secure it. Later deliveries would not need to be so carefully planted, for these would become enmeshed in the growing tangle of twigs. Finally, when the winter food cache neared completion, the beavers would bring heavy logs and place these on top of the underwater food store to weight it down.

Even as I watched Lily, I was conscious that other members of the colony, who had gathered below my viewing rock, were rapidly whittling away at the pile of branches she alone was making efforts to store. As luck would have it, I had brought an unusually big bundle of aspen that night, for John, who was visiting from Massachusetts, had come with me to the pond, and I was eager that he get a close-up view of every member of my colony.

Lily, as if aware that the others were rapidly consuming the resource she was in the process of stockpiling, returned quickly after each delivery to pick up and transport yet another branch across the pond. She worked at a pace that is unusual for a beaver,

Lily tows a food branch to the winter food cache.

for despite the descriptive epithet often applied to the species, beavers are not "busy" animals. On the contrary, they normally proceed at a leisurely pace, unburdened by outside pressures. One stick at a time they drag up on their house, one load of mud at a time they push onto their dam. And after doing a certain amount of work, they take a break to feed or groom or play or just float about in the water. But on this night Lily did not permit herself time off. She even forsook her evening meal and did not so much as sample the relished aspen she was transporting from one side of the pond to the other. On each return trip, however, she did pluck a lily pad and eat it on the move. I wondered: were none of the other beavers sensitive to the factors that had awakened in Lily such a compelling drive to cache food?

That Lily was in a race with the five hungry beavers who were fast devouring the food she had earmarked for storage seemed undeniable. On her final run, she collected eight of the largest remaining branches from under their noses and, taking care to add each one by the butt end to her already chock-full mouth, she swam off with the entire load. In moving this huge and awkward bouquet of scraggly limbs across the pond, she tilted her head to one side so as to elevate her cargo above the water and clear the lotus pads on either side of her. Had I not known that a beaver was the moving force beneath this apparition, I might have been quite baffled by the sight of an upright bush traveling across the pond on what appeared to be its own steam. And trailing after this sight were two kits, swimming hard and whimpering for sticks all the way: "uh, uh, Uh, UUH, UUUH! UUUH!" ("My stick! My stick! My stick!")

When this extraordinary spectacle came to a halt near the opposite shore, all three beavers, together with the traveling bush, went underwater simultaneously, Lily performing a most forceful dive to drive her load to the bottom. For as long as it took her to plant the eight branches, she and Lotus and Blossom remained out of sight. I wondered if the youngsters were taking part in the labor. I had my doubts, however, for I had read that kits do not help stockpile food until their second fall, by which time they are year-lings. Moreover, Lotus and Blossom were unusually young, having been born so late in the year.

John and I spent the next morning looking about the pond for evidence that the beavers had begun felling trees to be put into their winter cache. To our amazement, one swamp maple had evidently been taste-tested and abandoned. And to our delight, another one had been completely severed. As bad luck would have it, however, that tree had toppled against a tall oak and become entangled in it. And so the lumberjack who cut it received no satisfaction for his efforts. John struggled to pull down the hung-up tree and, after considerable tugging and by repeatedly slamming it against the oak that held it captive, he did at last succeed in disengaging the two. The swamp maple fell part way into the water, where we hoped it would be found and added to the meager store of branches Lily had planted.

We roughly estimated that to feed six beavers over a dozen weeks of winter might require a dozen large-crowned trees, having trunks measuring four to eight inches in diameter, be cut and stored. In addition, some two hundred smaller items—alder poles, willow brush, and saplings of many varieties—would need to be stockpiled. Although the region was heavily forested, most of the tree stands consisted of aging oak and swamp maple, interspersed with thickets of mountain laurel. Even though beavers eat these species, they do not favor them, nor do these kinds of trees regenerate quickly. By contrast, the plant species beavers most like, namely every kind of water-loving willow, black and yellow birch, alder, fruit trees, and above all poplar (every variety of poplar—cottonwood, big-toothed aspen, quaking aspen, and black aspen) are fast growing and resilient, capable of sustaining heavy use. Poplar, upon being cut, sends forth clones from its root system, perfect copies of itself, and even its cut stump will sprout a new tree. It also reproduces by the slower method of seeding.

But even such prolific species as these cannot renew themselves indefinitely. When exhausted by repeated cutting, they die out. It is then that beavers begin to bite into hard oak and swamp maple and will even cut and store a certain number of conifers. At this point in the evolution of a pond, most colonies abandon their beautiful waterworks and seek a new site to transform to their liking. If, however, the region is dense with beavers or has been converted by human beings to such uses as are incompatible with beaver life, an impoverished colony may be forced to remain in place and make do on less than ideal fare. Sometimes such colonies fail to store enough winter food to last until spring and, as a consequence, die inside their lodges.

Now John and I found several freshly cut laurel stumps and some tooth-etched barberry bush sticks and realized that the day of reckoning was not far distant for Lily and her family. This was indeed poor fare for beavers. I wondered what they would find to eat even before the pond froze over, after the first cold snap caused the lily pads to wither.

That night we returned to the pond and placed another big bundle of aspen branches in the cove. The yearlings, Laurel and the Skipper, began feeding on our offering as soon as they discovered it and they were soon joined by the kits. Then Lily appeared and, without stopping to sample even one of the yellow leaves that still clung to some of the branches, she began hauling away her find. As she had done on the previous night, she gathered up as many as she could carry in her mouth before starting for the other side.

Meanwhile my attention was attracted by sounds of splashing at the dam-end of the pond, where that morning John had liberated the swamp maple from the grip of the oak. Even in the dying light, I was able to make out a beaver, obviously the Inspector General, struggling to drag the big tree the rest of the way into the water. His every heave was accompanied by a loud grunt, and recalling the tree's weight, I fully sympathized with him. Even so, his vocalizations made me laugh. He sounded like a human being, pushing himself to the limit.

When at last he succeeded in pulling the hardwood into the pond, things went easier for him. Once buoyed, it became manageable and could be towed along the donut ring to the family lodge. The big beaver set his right shoulder hard against the trunk, which he steadied with his forepaws and gripped with his tong-like teeth, then set off. In the dim light, the traveling tree looked and moved along the far shore like an overloaded barge being navigated with extreme care. What on earth was this beaver going to do with such a cumbersome tree once he reached his destination? Surely he could not plunge it into an underwater food cache.

The answer to my question was soon forthcoming. When in front of the lodge, he stopped and began clipping branches from its crown, and these he scuttled, one by one, as he shifted back and forth from a towing modality into that of cutting, then diving and stowing. As I watched, I could not help but be impressed by the number and kinds of tasks involved in the business of laying in a winter food store. What was taking place seemed too complex to be casually dismissed with the simplistic word "instinct." Although cutting, diving, and storing *motions* likely are wired into a beaver's brain, all of these stereotyped units of behavior must be enacted in ever changing sequences to produce the desired *outcome*. Surely there is an organizer behind such activity. "Instinctive behavior" is a useful term, but its ready application to every animal act too often forecloses further inquiry as to how animals use their minds and suggests to many people that they do not.

As much cause for wonder was the fact that this beaver hauled the large tree to the site where he wanted it before cutting it into pieces. By so doing, he spared himself innumerable trips from lodge to fallen tree and back again. But what prompted him to do this?

Clearly, he was not following the path of least resistance. On the face of it, cutting the branches in place and towing them one at a time would seem the less demanding procedure. How amazing that an animal would elect to overexert himself at an immediate task (hauling an intact tree for a considerable distance), and so achieve a long range advantage (storing food more efficiently). Just as Lily saved herself time and work by gathering a bundle of branches in her mouth before crossing the pond, so the Inspector General got the most food planted in the least time by first performing what seemed a Herculean task. If he did so without any insight into what he was doing, then nature has indeed been generous in encoding this species with cost-effective responses.

Even as we watched the Inspector General plant swamp maple in the food cache, Lily traveled back and forth across the pond, transporting what remained of the aspen branches we had brought. When all were gone, John and I broke off boughs from an aged apple tree. These we placed in the water, where Lily soon found and approved them for storage. Predictably, she set off with several in her mouth, holding her bundle high out of the water. Even so, one snagged in the lily pads and was pulled from her mouth. To my delight, it was picked up by a trailing kit, who then struggled to tow it to the opposite shore. But the baby beaver had grasped it not by its stem end, but by its scraggly middle, and for this reason his progress was seriously impeded. The wrongly held branch rode in front of him, creating water resistance, which caused him to swim in zigzags. I shined my light on him as he collided with a tangle of lilies first on one side of the channel and then on the other. Yet he persisted until at last he reached the other side and, once there, he made a valiant effort to sink his contribution in the food cache. The unwieldy branch, however, defied him and bobbed to the surface.

The next evening this kit had hardly begun to dine on the single piece of aspen I laid out when his mother appeared and took possession of it. Once again he trailed after her as she crossed the pond, and once again he went underwater with her as she planted the appropriated branch in the food cache. It seemed clear to me that, with or without any intention on her part, Lily was serving as a role model for her offspring. Inadvertently or advertently, she was demonstrating the proper way to tow and scuttle food. I wondered if the youngster, who had had so much difficulty transporting the apple branch on the previous night, would profit by her example.

While the two beavers were so occupied on the far shore, I quickly laid out another aspen branch and awaited their return. To my delight, the kit swam back alone. I expected he would begin to gnaw on my offering, but he surprised me. Instead of feeding on it, he swam around and around it until satisfied that he was positioned to seize it by the butt end. Then he began towing it across the pond. I felt like cheering when I saw how easily it floated to his side, hardly interfering at all with his progress. And when he arrived at the lodge, he managed to sink this rightly held contribution without much difficulty.

Had he learned the proper way to transport and sink a branch by watching his mother? Had he hit on the solution through trial and error? Or was he simply programmed to perform the task correctly at a certain stage in his development?

I do not have any definitive answers to these questions, but I do have opinions. As far as I am aware, kits as young as the one I watched have not been observed to lay in winter food, and so, even if this youngster would have automatically manifested this behavior in due time, I think it unlikely he had reached that stage in his development. Whether the youngster got the hang of towing branches the right way round by watching his mother or by trial and error is, of course, a judgment call. But since I did not see him make any random trials (he went directly to the solution after having seen his mother do it correctly), I am of the opinion he learned by imitation. It is possible, however, that the real answer to my questions is "all of the above." Perhaps such things cannot be teased apart.

Chapter Eight

I n November the beavers advanced their clocks and emerged
from their lodge at four in the afternoon and sometimes even
earlier. Perhaps they needed some brief exposure to the sun's
ultraviolet rays, which they would not have received had they held
to their usual routine, for daylight was shrinking with every passing
day. Did the late-born kits, whose bones were at an unusually early
stage of development for the season, crave a dose of vitamin D?
And did they arouse the entire colony as they climbed over sleeping
bodies on their way to the plunge hole? Whatever the reason for
the forward shift in their activity cycle, I was delighted by it. It
gave me some illuminated time in which to view them.

The colony's food cache had now become visible; tops of stored
branches protruded above water. From what was showing, however, I had to conclude that their winter food larder left much to
be desired. High bush blueberry and leathery-leafed mountain laurel appeared to be the only vegetation the beavers had collected
during the week I had missed going to the pond.

Interestingly enough, I had learned something about storing tree
parts during that interim. While visiting John in Massachusetts,
the two of us had stacked branches of a tree he had cut, and we
were both amazed by the heap of brush they created.

"Let's take a page out of the beavers' book," John finally suggested; and we pulled the tangled, billowing mass apart and started
over. This time we lined up a number of boughs so that all their
butt ends pointed in the same direction. Then we inserted the

remaining branches into this twiggy structure, just as one slips extra flowers into an already arranged bouquet. The result was a neat and compact stack.

"They even got that right," John commented.

Now I viewed the top of the beavers' underwater food cache with new respect, albeit some dismay. Would they survive on such stuff? Why hadn't they moved? Was it because the kits had been born so late in the year? Whatever the reason, they would not leave now. Not enough time remained for them to convert a new site into a pond, build a lodge, and cut and store a sufficient number of trees to carry them through the fast-approaching winter. Already a thin layer of ice formed on the pond at night, and though it melted during the heat of the day one cold snap would change that.

By now all the lilies had died and the beavers were surviving on barberry bush, a plant so prickly I couldn't handle it. The debarked stems of that garden escapee were bright yellow and so could not be confused with those of any other plant. They littered the shore. I had expected that the beavers would have cut another swamp maple by this time or perhaps even felled an oak, but no tree work had been done during my absence. Dan Pierson, after viewing the situation, suggested we give the colony a hand, haul in some cut poplar for them to store. "I've located a big stand on our property that they can have," he offered.

I was torn. My head told me to leave the beavers to their fate, whatever that might be. Nature is always right, even when cruel. If animals make some fatal mistake—fail to lay in winter stores— that particular tendency to be shortsighted will be reduced in the population with their demise. I told myself that mortality is a natural part of existence, beavers must die to make room for future generations, that the catastrophe about to befall my colony would keep the species in balance. I told myself that here in the park where animals are not artificially reduced by man through hunting or trapping, numbers ought not be artificially bolstered by man through feeding. The park beavers were doing a fine job of regulating their overall population, and I ought not interfere with that.

It was cold sitting by the pond. The darkness I had grown to enjoy was once again becoming hard for me to endure. On most nights the sky was overcast, and so I enjoyed neither light from the moon nor a view of the cosmos to make my long vigils more tolerable. By now the hum of insects, the shriek of tree frogs, and every other cheerful life-sound had ceased. Only the screech of an owl occasionally brightened the cold, dark hours I spent at the pond.

One night I heard the report of a gun at close range and I knew that deer poachers were close by. After that, my fear of being alone in the park after dark returned; in fact, my imagination worked overtime creating worst-case scenarios. Every rustle I heard was a human footfall. At times I envisaged I was being sighted in the cross wires of someone's gun. Was I to be mistaken for some poor beast and shot? And what kind of person would poach deer anyway? I didn't want to meet up with such an individual in any circumstances, let alone in this dark and lonely place. I shut off my lantern and sat quietly in the blackness.

The next day I called Kenneth Didion, the local game warden, to report the shots I had heard. Since Didion's district included all of Rockland and part of Orange counties, he was a busy man. And now that deer season had opened, he had to be everywhere at once. Clearly it was not a good week for him to stake out a state park where hunting is not permitted. Nevertheless, a few nights later he stopped by the pond to let me know that he had caught two deer poachers red-handed, not far from Lily Pond. I felt great relief at this news and hoped it would put an end to the gunfire.

Didion had recently been transferred to Rockland County from the Adirondack Mountains, some two hundred miles north, where poplar is abundant and conditions are generally more favorable for beavers. He told me how much he had enjoyed watching beavers there and he expressed an interest in my colony.

"They make a hard living here," he commented. "It doesn't look like they have much to eat. I wonder if they'll make it through winter."

He then assured me that he would make intermittent checks for illegal traps, which might be set at the pond now that the beavers' fur was in prime condition. And he described how such drowning devices work, so that I, too, could keep an eye out for them. The trap itself, he explained, would be placed in shallow water near to the shore. Should a beaver step on one, it would snap shut on the animal's foot, and at the same time the entire works would slide down a long guide wire, one end of which would be staked in deeper water. There the beaver would be held underwater and die of drowning. (Not quickly, of course. Beavers can do without air for as long as half an hour, during which time some victims struggle so hard to free themselves that they wring off their trapped foot.)

After my conversation with the warden I checked the pond every night, looking for stakes and wires. Soon ice and snow would conceal these hard-to-find sets, and only the poacher who put them in place would know their precise location. He, of course, would return to collect his illegal booty and reset his traps when no one was around to observe him.

One cold evening I arrived at the pond and found that most of it had frozen. A single navigable channel led from the lodge on the south shore one hundred yards to the dam on the west end of the pond, hugged that long structure for fifty yards, then turned and followed part way along the north bank. What I had called "the donut ring" had now become a crescent. In addition, a small pool of open water directly in front of the lodge continued to serve the kits' need to dive and porpoise, an activity they happened to be engaged in as I arrived. I suspected that heavy beaver use in these areas had kept the water roiled and so forestalled its freezing.

For several minutes I sat and watched the kits play before spotting the Inspector General. He materialized from nowhere, simply bobbed up from under the ice into the open water alongside the dam. After checking that engineering work, he dove and vanished.

Where was he headed under a roof of ice? How did he find his way about now that he could no longer surface to take his bearings? Did he carry an underwater map in his brain? And how could he be certain of finding a vent hole, a place to lift his head and inhale a breath of air, should he suddenly need oxygen? Do beavers strategize about such matters? Do they possess some kind of internal alarm mechanism to warn them when their oxygen supply is getting low? "YOU HAVE FIVE MINUTES OF OXYGEN LEFT. DO NOT DELAY! GO AT ONCE TO THE NEAREST VENT HOLE WHILE YOU CAN STILL MAKE IT!"

I expected him to show up somewhere along the crescent, but he surprised me and swam a very long way under the ice, a good football-field length, before surfacing in front of the lodge. There, after touching noses with the kits in greeting, he went to work, breaking off slabs of ice all along the edge of their swimming hole. He used several techniques to do this. In places, he pressed down on the ice with his front paws until a slab broke off and floated away. Where it was too thick to be dismantled by this method, he pulled himself on top of it and let his body weight cause a section to split off. He also employed another technique: he swam underneath the ice and bumped it from below. It was great fun watching him do this. One hard thud and the ice would crack. Then on a second or third ram from below it shattered, and the Inspector General would pop right through, like a plump show-girl bursting out of a cake.

I was impressed by this behavior. Whether or not he knew it, by keeping water routes open, this beaver was postponing his consignment to a subaqueous realm—an unlit, claustral world soon to be endured without the palliative of a long winter's sleep. He was also buying time in which to add more branches to his paltry food cache. Was he able to anticipate what was about to befall him and was he trying to forestall it? Had he some memory of past winters? Can beavers identify a future effect (in this case, long confinement) with an immediate phenomenon (in this case, the formation of ice)?

Of course, there is a theoretical explanation for this beaver's behavior, though it does not attempt to probe what subjective thoughts and feelings an animal might experience while carrying out such an activity. Breaking ice must have had survival value for beavers. Over the species' long evolutionary history, those individuals inclined to break ice probably succeeded in storing more food and endured fewer days of confinement than did those beavers not so inclined. Thus icebreaking beavers had an edge on non-icebreakers. They lived longer and therefore had time to produce more young, who, it can be supposed, inherited their peculiar tendency to break ice. In time, icebreaking individuals swamped the population and the trait became coded in every beaver.

But is it? Do all beavers break ice? Perhaps so, though I had never heard it reported. And how had any beaver hit on this behavior in the first place? Was some mutant individual born with a compulsion to break ice and did he do so with no particular objective in mind? Or was there some dimly perceived purposefulness behind his behavior?

Such questions cannot be answered by science, for it is impossible to investigate and quantify that which is experienced subjectively by another, be it man or animal. Yet in deliberately excluding such questions from their purview, scientists unwittingly leave a good many people with the impression that nothing goes on in an animal's mind, that animals do not experience emotions or process thoughts, that they don't even register sensation to a degree that should concern those who inflict pain on them. Such people overlook the fact that experimental psychologists would not be able to condition laboratory animals by means of electric shock were their subjects unresponsive to pain. Sensitivity to pain, however unwished for, has long served animal survival, as it has our own. The same must be true for emotions and memory and cognition and even reasoning. Man did not spring into being in full possession of those faculties, which he likes to identify as uniquely human, but slowly acquired them as he made his way up the same phylogenetic tree ascended by all other mammals. During a very long evolutionary process, man and animals have had to pass many of the same survival tests and have developed similar equipment with which to do so. It is a specious (and species-ist) conceit to imagine that no demand has been put on animals to make use of their brains and to experience the functioning of their nervous systems.

Over the next few nights the pond ice grew thicker and open water diminished, despite the Inspector General's tireless efforts to prevent this from happening. Now I saw that he had been joined by Lily, who seemed equally determined to stave off incarceration, though as an icebreaker she was not the underwater rammer that he was. Lily's preferred method of operation was to heave herself atop the ice, whereupon she would perform a kind of jiggling dance until the edge upon which she perched gave way and sank beneath her.

When darkness blotted out the scene, I became more sensitive to the strange acoustical effects created not only by the beavers' war on ice, but by the freezing process itself. Once I heard a sound like that of metal on metal and was certain that Lily or the Inspector General had stepped into a trap and been swept down a long guide wire to a watery death. My single thought was to find and rescue the animal before it was too late. But this was no easy mission. I was afraid to cross the pond on new ice, and it took me some time to make my way around its slippery shores in the dark. Then after experiencing much difficulty crossing the dam, I ran into an impenetrable barrier on the far side. The dense thicket of mountain laurel I had so admired in the spring protected the beaver house from approach by land, and I was forced to turn back.

The next night I arrived to find both the crescent and the kits' "swimming hole" had frozen. I would not get a head count of my beavers, for they were now sealed under ice. I hoped all six were safely sleeping in the high and dry chamber of their snug lodge. A vent hole at its apex would admit what air they needed, though no light would penetrate the structure's thick walls. The beavers were consigned to darkness now.

I wondered what might be going on inside their dark chamber. Perhaps the kits had awakened and slipped out to clip a branch from the food cache. Or perhaps they were enjoying a long swim under the pond's glassy roof. How strange it must seem to them suddenly to find a lid on their world. Would the youngsters forget it was there, attempt to surface, and bump their heads?

I remained at the pond only a few minutes. There was no point in staying longer. I did not give up hope, however, that I would see my colony again soon. It was early December and I expected at least one thaw would occur, if not by Christmas, then perhaps in January. For that matter, the winter might prove mild, and I might find myself beside the pond watching beavers swim about in open water at various times throughout the season.

Meanwhile, I put the troubling, screeching sound out of my mind and bade my beaver friends farewell and safe wintering.

Chapter Nine

L ily Pond had become a blinding Arctic snowscape, a sweep
of white inscribed with animal tracks that meandered about
its glistening surface like the loose handwriting of a giant.
The dense border of mountain laurel along the south shore would,
of course, remain green all winter and overarch the beaver lodge,
providing it a measure of concealment.

John and I stepped off the north bank onto what had only a week
earlier been open water and headed directly across the frozen pond
toward the beavers' snow-domed quarters. In our down jackets,
wearing heavy socks, and sensibly booted, we were dressed to spend
a few sunny hours out-of-doors in the freezing weather. Afterward
we would sit by a fire and warm ourselves with hot food and drink.
What an advantage we human beings have over wild animals, I
thought. The doe who regularly watered at the pond, the fledgling
owl whose squacks had startled me, the fox who frequently left his
calling card near my viewing rock, the mouse who sometimes ran
over my feet, these animals would gain no relief from the cold now.
Everywhere their movements would be impeded by snow, and they
would have difficulty making a living. Soon they would become
vulnerable to all causes of mortality.

December, of course, was only the opening act in this annual
drama. As the cold season progressed, wildlife would become increasingly stressed. Inadequate nutrition would reduce their speed
and efficiency, making it ever less probable that any food-deprived
animal would obtain another meal. Bitter nights would drain energy

reserves. Before spring, large numbers of every kind of species would fail this harshest of nature's tests. For winter is a merciless leveler; it ravages hunter and hunted alike, rendering moot the question of what eats what. Winter's voracious appetite leaves only the strong to beget young.

I did not despair for my animal friends, however. There are winters and winters. Some years are kinder to wild things than are others. Moreover, certain species have devised ways of coping with the season's exigencies. For example, eons ago bats and bears and woodchucks discovered the trick of hibernation. By turning down their metabolic rheostats, they are able to sleep through the cold weather and so have dispensed with the necessity of searching for food. Other species, such as the vole and the squirrel and the beaver, hit upon another strategy. They prepare for winter in advance. They minimize the impact of cold weather on their energy reserves by constructing warm living quarters and stocking these shelters with feed. Yet some species make no provision at all to meet the vicissitudes of arctic weather. To stay alive, deer and fox and birds of prey must continue to forage or hunt on a daily basis. Winter is most difficult for such animals.

Now that we could walk on the frozen pond, John and I made a beeline for the beavers' lodge to inspect it close-up and from every angle. As we neared the snow-domed fortress, we were startled to see vapor curling out of a vent hole at the apex of the structure, a phenomenon that gave the place the appearance of a miniature igloo inside which little Eskimos had built a fire. In reality, the "smoke" we were seeing was condensation of the animals' warm breath escaping through the vent hole and meeting cold, outside air. Or perhaps a wet and steamy beaver, having just entered the living chamber after an under-ice swim, had displaced warm air and sent it curling out the lodge "chimney."

We explored a number of such possible explanations, none of which appealed to me as much as the mental picture I had formed of fur-clad Eskimos gathered about a fire. And that image became hard to shake when a hum of "beaver talk" commenced from within, the very tones and inflections of which sounded like human voices. I suddenly understood why, in times past, certain Native American tribes viewed beavers as "little Indians" and did not kill them. Not only did the animal play an important role in tribal mythology, but sometimes a kit was brought to camp and kept as a pet. The Crees, for example, saw fit to present any tribeswoman who lost her infant with a baby beaver as a means of easing her bereavement. Sad to say, this sympathetic view of *Castor canadensis* did not survive the arrival of Europeans to North America. Even the Crees were eventually persuaded to act as guides for French and British trappers, whose quest for beaver fur was all but insatiable.

"Well, the beavers in that lodge certainly sound healthy," John commented, as we stood listening to their murmurings.

This seemed something of a miracle in view of the poor quality and the paucity of the food they had planted. A few reddish blueberry twigs protruded above the ice where a haystack of tree branches should have revealed its tangled top.

"Maybe they've dug such a deep storage hole that only a little of what they've actually put into it shows," I ventured, hoping to hear an encouraging response from John. None, however, was forthcoming.

Now that snow blanketed the ice, we were no longer able to peer through it to determine how much and what kinds of food had, in fact, been stored. Nor were we able to look for "beaver bubbles," created when swimming animals expel air from their lungs. Like helium balloons, these globules of carbon dioxide rise to the surface until they become trapped under the icecap. Later, as the pond continues to freeze, they become incorporated in the ever thickening mass. Not all ice bubbles one sees, however, are caused by beaver breath. Some form when trapped air escapes from the animal's dense coat, which is frequently groomed and fluffed up inside the ventilated lodge.

The presence of ice bubbles is a useful sign to beaver watchers and indicates that a colony is wintering successfully. Ice bubbles also serve beavers. Conceivably, a string of bubbles could serve as a source of oxygen to an animal who has traveled far from his lodge and suddenly experiences the need to breathe. By pressing his nose into them, such a hapless beaver might obtain enough air to make it to another, better source. More plausibly, bubble-perforated ice offers the beaver a weakened site in the ice cover, a place that can be punched out from below. Thus wintering colonies do gain access to open water and fresh air. Since such honeycombed ice is likely to give way under the weight of a human being, one must take care not to walk too near a beaver house in winter, no matter how low the temperature.

When not blanketed by snow, pond ice serves as a kind of map by which one can trace the underwater movements of wintering beavers. By following a string of bubbles, for example, one is likely to be led to the main dam, which is visited and maintained by beavers even in winter. Any under-ice leak is plugged with bottom muck to assure sufficient water level is maintained to cover the lodge's exit-entry holes. Conversely, beavers sometimes lower water levels by gnawing a hole in the dam. As water siphons from the pond, a sandwich of air is created between the surface of the pond and its ice cap, and this provides the resident animals a ready supply of oxygen. Moreover, the flow of water through this beaver-made outlet weakens the ice alongside the dam, making it easy for them to punch through and go up on land.

When you think about it, life under ice is an extraordinary adaptation. By wintering in a fairly constant micro-environment, the beaver guarantees itself above-freezing temperatures (that of unfrozen water). Even so, *Castor canadensis* seems surprisingly well equipped to withstand the effects of extreme cold—at least for limited periods. Obviously, the animal has been provided with a warm coat; but what of its unprotected feet and tail? I have watched beavers travel over deep drifts, their naked paws sinking deep into cold snow. And what of the hairless tail that drags across these tracks and rubs them out? How is it that the beaver's unprotected parts don't freeze when my own well-shod feet feel like painful stumps as I stand and study evidence of the animal's above-ice activity?

Certainly, beavers must pay attention to weather conditions before venturing up on land to forage. For should temperatures plummet while they are out, the opening through which they exited could freeze shut, locking them out of their pond. Thus cut off from warm lodge and food cache, such an unfortunate animal would be doomed. Moreover, in exchanging a 33-degree environment for one that is much colder, the animals could incur an energy deficit, no matter how much good feed they ingest on shore. Surely beavers

A beaver has discovered open water and enjoys a brief respite from winter confinement.

must possess some kind of weather antennae that prevent them from making faulty forecasts. In a study done at the Quabbin Reservoir in Massachusetts, Richard Lancia, Wendell Dodge, and Joseph Larson found no sign that any of their subjects ventured from their ponds when thermometer readings were 14 degrees Fahrenheit or below. As temperatures went over that mark, however, above-ice activity increased proportionally.

Still I once came upon evidence that beavers had emerged from their pond on a night when the wind-chill factor was something fierce. Moreover, on that bone-chilling morning, I found such a maze of big and little tracks around a newly felled oak that I had to conclude the tree-cutting event had been well attended. Moreover, the various-sized tooth gouges etched in the stump and fallen trunk suggested that no single beaver had done all the work. Perhaps, in this case, no more energy was spent than absorbed, since many had shared in the labor, and all had partaken of the feast.

Winter tree-felling by beavers can produce some rather mystifying results. If, for example, an animal perches on a deep drift while making his cut, the stump he leaves behind may be as tall as a man. After the snow melts, this curious relic is likely to give rise to rumors that giant beavers exist in the area. One such stump, measuring eight feet and eight inches tall, was discovered near the Lewis River in Montana and created quite a stir until a local woodsman explained how it had happened.

Since the Lily Pond beavers were not inclined to cut trees at any time of the year, John and I did not anticipate finding evidence of such activity on this morning in early December. Still there was plenty for us to look at. We ducked under the drooping laurel and crawled onto the lodge to inspect the air inlet at its apex. Frost had formed around its edges—a healthy sign that the beavers inside were warm and breathing. I had hoped to make use of this opening as a peephole, but my idea turned out to be naive. Even had the interior of the beavers' house been illuminated, I would not have been able to look through the air vent, which, as it turned out, was not a hole at all, but a lattice of loosely crossed sticks—a gridwork that the animals had left unplastered when they had mud-slathered the rest of the lodge's exterior.

Now the previous night's snowfall had added still another layer of insulation to the big beaver house and, with six animals generating body heat inside, their living quarters must have been warm indeed. Moreover, Lily, the Inspector General, Laurel, the Skipper, Blossom, and Lotus, were probably huddling together, for living space within the huge lodges that beavers make is always limited. The circular floor plan upon which every member of the colony must seek a space often measures no more than four feet in diameter. And a low vaulted ceiling may even prohibit large occupants from rising up on their hind legs.

I had learned something about the interior of beaver houses some weeks earlier by wading out to examine an abandoned island lodge* situated in the middle of a partially drained pond. What mud had covered that once impressive residence had completely washed away, enabling me to insert a yardstick here and there through its wicker-woven walls. Thus I was able to take measurements of all the open spaces within. Of great interest to me was the discovery of a small

---

*The Lily Pond beavers inhabited a bank lodge, which extended out from the south shore like a peninsula.

antechamber at a level slightly below and leading up into the animals' living quarters. Attached to this antechamber was a kind of chute that had once opened underwater and had obviously served as the animals' main entry-exit ramp. Probably the antechamber itself was used by wet beavers as a place to drain before they re-entered the family's high and dry living quarters—like the vestibule where we drop wet galoshes and stow our dripping umbrellas. The lodge boasted two additional entry-exit tunnels, but these were not so wide as the first, nor were any antechambers built into them. Perhaps they were used primarily as exit ramps.

A loud glug brought my attention back to Lily Pond. One of the beavers had evidently emerged from the lodge and was moving about somewhere under the ice—probably visiting the food cache. We strained our ears trying to track his movements and managed to pick up sounds of gnawing at the animals' underwater pantry. A moment later another glug, and then murmurings from within the lodge, told us that he had returned to warm dry quarters.

I tried to imagine what it would be like to winter with one's closest relatives in a totally dark, relatively airless, and overcrowded room. How do beavers keep from getting on one another's nerves? Rats become vicious when crowded. Even the sociable wolf, a species that has evolved impressive strategies for maintaining good pack-relations, likely would lack the social skills necessary to survive such prolonged intimacy. And many human beings who have had to endure long confinement with others have afterward sought to explain their bad behavior as the effects of "cabin fever."

It was becoming ever more clear to me how underreported is the innate good nature and sociability of the beaver. And no wonder. What a casual onlooker is most apt to see of the animal—colony members emerging singly from the lodge and heading off in different directions to feed—does not suggest that it is gregarious. For months I had taken this pattern of behavior to mean that my beavers were relatively aloof. But now, as I reflected on lodge life, I had to ask myself why I should expect a group of animals who have been confined together for a twelve-hour stretch to rush outside and greet one another. How much intimacy can even the most social of species express?

Looking at it from that point of view, I was suddenly impressed by the amount of interaction I had observed. For not a single night had passed but I had watched my beavers swim alongside one another, or touch noses, or "speak" to one another. And how often had I seen two or more of them seek one another out for no other reason than to feed in company? Yet the lilies they consumed side by side were available all over the pond. And what of the precision diving and porpoising bouts I had witnessed? Were these not expressions of exuberant play between friendly animals? Moreover, what few conflicts had erupted between any of my beavers had always been resolved by "verbal" means or by one beaver rushing at the other or by a push match during which no injuries were inflicted. Never had I seen one beaver actually attack another.

But the most impressive evidence of sociability I had noted in more than six months of beaver watching related to the caregiving behavior of nonparent beavers toward the young. Laurel and the Skipper had lavished attention on the infant kits. And even after the youngsters had matured to the point where they were relatively self reliant, still they would sometimes be indulged and allowed to climb on a back or to share a stick.

All this suggested to me that *Castor canadensis* might well prove to be one of the *most* social of mammals. And now it struck me that what was going on inside the lodge right now in winter would be more revealing of the beaver's true nature than anything I could hope to see outside. I decided to pay occasional visits to Lily Pond throughout the season to try to deduce what life in the beaver house was like. I realized, however, that any information I would obtain would have to be gleaned entirely through my ears.

Chapter Ten

I saw my beavers two more times in December, and Dan was the reason. He persuaded me that we ought to cut a load of aspen and lay it on the ice, so that any beaver who made it through winter would find starter feed in the spring. Dan had consulted an expert at the University of Massachusetts who told him that beavers face the greatest risk of starvation not in winter, but just after ice-out and before new growth becomes available. I didn't see how this applied to the Lily Pond beavers, inasmuch as they had almost no food stored to see them through the winter and might not live until spring. On the other hand, a midwinter thaw might occur, during which time the beavers could stash whatever we would bring them.

While I wasn't convinced that feeding beavers was the right thing to do, it was clear to me that nothing at all would be proved by letting them die. It is a given that beavers who fail to prepare for winter do not survive, and I had much to learn from the animals I had come to know and identify at great cost to myself. So in the end, even though delivering a load of spring starter-feed seemed to be a case of doing too little too late, making that effort felt better than taking no action at all.

Therefore, Dan and his daughter, Nina, and John and I spent an afternoon cutting down small trees, which we then sectioned and bundled off to the pond. The day we picked to do this was sunny, and the ice around the beaver house was pocked with bubbles and hazardous. We dared not approach closer to the lodge than seventy-five feet. Moreover, in making each delivery we took care to space ourselves some distance apart, so as to distribute our weight across a wide area of the ice cover. Thus we toted and dragged bundles of branches and sawed-up tree trunks as far as we dared go, and when the ice beneath our feet began to creak, we stopped and bowled or slid our offerings toward a common heap we were creating just twenty feet in front of the lodge.

By this time the snow, which two weeks earlier had blanketed the pond, had either blown away or evaporated, and the glassy surface upon which we trod was wet and treacherously slippery. I carried a forked stick to support my every step. I also used it as a shuffleboard pole to shove my deliveries to their ultimate destination.

We had nearly emptied the van of sticks and trunk sections and branching twigs (enough feed, I calculated, to sustain six beavers for two or three weeks), when suddenly a resounding "whoom" rang out and the ice supporting our heap of aspen gave way. At the time, John and I were making our last deliveries, and for a moment, we expected the cave-in to create fracture lines and cause the very ice we were standing on to give way. But that did not happen. When the echo of that unexpected event died away, we gazed in amazement at the result: the tangled heap of cut aspen that we had brought to serve as spring feed for the beavers was afloat. And within one minute a furry face surfaced alongside it and stared long and hard at us. It was Lily.

I spoke to her. "We've brought you Christmas dinner," I said.

At the sound of my voice she dived, and a moment later the floating branches and sectioned tree boles began to jiggle and shift about as she worked to extricate a single piece. This she did not eat, but promptly dragged under the ice, apparently transporting it to the food cache. In a few moments she returned to pry loose another branch, which she also towed under the ice to the deep place the beavers had excavated to hold their winter food.

Then the kits appeared. Their beady-eyed heads popped up beside the floating aspen and they gazed long and hard, first at the wintry scene and then at us. But for their paddle-shaped tails, they might have been a pair of seals, peering out of an Arctic seascape. Then, as if by prearranged signal, those tails flipped into the air, and the two performed a somersault dive in perfect unison. Not long afterward we heard the sound of gnawing beneath the floating woodpile, and a long branch began to disappear. When it was no longer visible, a stream of under-ice bubbles leading toward the lodge told us that it too was destined for storage. The kits were helping.

While Dan and Nina waited at the dam, John and I stood on the ice and watched for half an hour as Lily and her late-born offspring, Blossom and Lotus, traveled back and forth towing branches from aspen heap to food cache. At four-thirty, when fading light and snowflakes made further viewing impossible, we made our way back across the ice. A wind had come up and the temperature was falling. We expected that by morning the hole, which had opened so magically to admit a fortnight's worth of food for the beavers, would seal itself shut again. We were cold and tired and hungry from our exertions, more than a little pleased over the unexpected result and eagerly looking forward to a good dinner ourselves. Back at the van, however, we were confronted with one piece of unfinished business—a large and still-intact tree we had put off sectioning. By now our strength and resolve had waned and nobody felt up to the task of cutting, bundling, and delivering this last offering to the beavers.

"Let's just drag it down to the pond and leave it for them to find in the spring," Dan proposed.

This suggestion met with no opposition. On the contrary, in one voice, we reminded each other that such had been our original intention anyway—to bring the colony some spring "starter feed." That we had gotten food to them through the icecap seemed something of a miracle. And so we carried the fifteen-foot aspen tree (whose sawed trunk measured three inches in diameter) down the bank and dropped it near the dam.

I didn't expect to see that tree move from the spot where we left it, but I did. Several nights later I visited the pond under a big and bright moon and spied a wet and glittering beaver up on the bank. It was the Inspector General and he was jerking and tugging on the aspen with all his might. The energy he was expending was only slightly less prodigious than was his determination to move his find off the bank and into a narrow slit of open water alongside the dam. This opening in the ice, caused by the action of water against that barrier or maybe even punched out by beavers, looked too small to receive the tree. Yet he persisted and eventually he triumphed. With much groaning, he slid the butt end of his prize into the slit in the ice and then, after going underwater himself, yanked the rest in after him. I could hardly believe my eyes as I watched the upper portion of that full-crowned tree gradually disappear from sight. When all grew quiet, I realized that the Inspector General was pulling it toward the food cache one hundred yards away.

I walked across the frozen pond to be on hand should the tree-towing beaver actually make it to his destination. En route I tracked the sound of his progress, followed the glugs and scrapes that his cargo produced as it bumped against the ice roof or dragged over a bottom boulder. As the two of us progressed by fits and starts, I hardly dared hope that this huge effort might succeed. Perhaps the hard-working animal would locate some deep channel that he and his fellow colony members had dredged for just such an event as this. But no such shipping lane existed to connect the beavers' food cache to the place where we had dropped the tree. And so the Inspector General labored on.

Not until he had put seventy-five yards behind him did I begin to believe the big patriarch was going to make it. At that point it seemed conceivable he would drop the tree, return to the lodge for air, then come back and finish up the job. He did not, however, do this. Nor did his efforts cease when at last he did arrive at the food cache. Instead, to my utter amazement, he began gnawing the tree into sections. What kind of lungs did this animal have anyway? How was it possible for him to continue exerting himself without first answering that most primary of physiological needs—inhaling oxygen?

I tried to picture what was happening beneath the ice. To gnaw branches underwater without choking, a beaver must first suck his loose cheeks into the sides of his mouth until they meet in a space between his back molars and his bright orange front incisors, the teeth that do all the work. These sucked-in cheeks form a kind of dentist's rubber dam that prevents water from getting into the animal's lungs while he whittles and clips with his four exposed, large, curved, and extraordinarily sharp front teeth.

After several minutes passed I received no more auditory information. The Inspector General, evidently in need of oxygen at last, had returned to his ventilated lodge to take a well-earned rest. Although I hadn't timed his underwater labors, I estimated that he had performed that strenuous work for at least twenty minutes without air.

For days after that I sat on the beaver lodge, listening for sounds of life within. To my delight, I was privy to a number of "conversations," mostly of the "uh, uh, uh" variety. I was also able to hear animals depart and return to the lodge. A hollow glug announced when one hit or emerged from the water. It soon became clear to me that the animals were clipping branches from the food cache and bringing them inside to eat. Moreover, when a beaver reentered the lodge a dialogue usually commenced, the tones of which suggested that two age-classes were participating. The kits' end of the conversation always sounded imploring. "Give me some," they seemed to be saying. And my ears told me the little wheedlers more often than not got what they wanted. For after all discussion ceased, I usually heard sounds of *two* beavers gnawing on bark.

To verify what I was hearing, I obtained a copy of a study by French Canadian researcher Françoise Patenaude. Ms. Patenaude built a viewing hut into the back of a beaver lodge in order to see what took place inside. What she reported agreed entirely with what I was deducing by ear: that the adults and yearlings made frequent trips to the food cache to obtain food for the kits. Thus the youngest and most vulnerable members of the colony were spared having to make a good many trips into icy water.

Other facets of lodge life, not so apparent to a mere listener, were also reported by Ms. Patenaude. Throughout their long winter confinement she frequently saw the beavers nibble and groom one another's fur. This pleasurable activity seemed to cement family bonds and promote harmony within the group. And when on rare occasions Ms. Patenaude witnessed confrontation between two beavers, she reported that the animals settled their differences by engaging in a harmless shoving match. No teeth were bared, no animal was bitten.

Ms. Patenaude's observations confirmed what I had long suspected: beaver families employ sophisticated strategies to keep the peace in the most trying of circumstances. Not only do they perform altruistic acts (bringing food to the youngest, weakest members of the colony), they also defuse aggressive feelings through harmless ritual (wrestling matches). The beaver is indeed a most social being.

Chapter Eleven

Snow fell several times during January, obscuring my view of life under ice cover and turning the beaver house into an igloo. I circled the shoreline looking for sign that some member of my colony had found an escape route from imprisonment, but saw no webbed tracks or tail-drag marks in the snow. I did discover other animal life, however. Wild turkeys were roosting in the dense laurel near the lodge. And otters had left a scat pile of fish scales beside a narrow opening in the ice, evidence of a recent feast. Afterward they had climbed up the bank just to coast down again on their bellies. A deep trough in the snow marked where they had made numerous such toboggan runs. Otters do that kind of thing just for the fun of it. My most delightful sighting, however, was a tiny saw-whet owl, perched like a beautiful Christmas ornament in a laurel bush that overhung the beaver lodge. But this bird was no hand-crafted work. Its feathers were too perfectly arranged to have been created by the most skilled of artisans. It showed no fear of me. I parted the branches six inches in front of its unblinking, yellow eyes, and it did not so much as shift its weight to ready itself for flight. Saw-whets are like that. Roger Tory Peterson describes the species as "ridiculously tame." Now this bird-jewel conjured up a memory of another little owl.

My brother had found it on the ground and carried it home in his handkerchief. It belonged to him, he informed me, but if I wanted to be useful, I could be its nurse. Within one day of becoming captive, my little patient took raw hamburger from my hand, and, within one week, it perched on my finger. My mother permitted the bird to be housed on our screened porch, which was not used in winter. There it could enjoy ample air space to exercise its wings. Nevertheless, my ignorance of bird physiology resulted in the little owl's demise. In retrospect, I believe the diet I fed it lacked calcium, but I knew nothing of such things at age ten.

Now I longed to press one finger against this bird's breast to see if it would be as accommodating as Oscar and climb onto my hand. But I resisted. Having once done mortal harm to just such an owl, albeit with the best of intentions, I did not want to reenact even a tiny part of that crime. I let the laurel branches close around it and went away.

During that month I drove up to northern Massachusetts to visit John, and there we watched beavers swimming under clear ice. They shot about like seals on a clockwise course around their pond with no apparent objective in mind except to experience their own motility. I was just amazed at their speed. When paddling on top of the water, beavers appear to be sluggish swimmers.

In mid-February Lily and the Inspector General mated. I didn't see it happen, but of course it did, for kits emerged from the lodge the following spring. In another year and in another place I would watch another pair engage in foreplay before consummating their union in open water. On that occasion, I guessed wrong about the sexes of the two and later had to revise my notes to agree with the facts. As it turned out, the smaller beaver was the male and he actively courted the female by playfully tugging at a stick she was gnawing on. His behavior, instead of arousing her ire, induced her to swim with him, and not just take an ordinary cruise around the pond either. Both animals gripped that single stick with their teeth and dived and surfaced and paddled about, all the while "vocalizing" in dulcet tones.

Beavers mate in water, normally during winter. This means that in places where February temperatures are below freezing, the animals somehow manage to copulate under ice. It should come as no surprise, therefore, to learn that not many people have witnessed this event. Those who have describe it as follows: the female floats, belly down, and allows the male to come alongside her and grip her torso with his forepaws. He then rotates the lower part of his body, twisting it under her so that the pair's cloacas meet. In this awkward position, copulation takes place. If the union, which lasts from thirty seconds to three minutes, is successful, kits are born 107 days later. If not, the female comes into heat again (as many as four times during the mating season), and the two have another go at it.

Some researchers believe that the male must always be the smaller of the pair, so that the female can support the two of them during the mating act. I disagree. The Inspector General was considerably larger than Lily, and there could be no doubt that he was the male and she the female. Not only was Lily's sexual identity clearly manifested in the way she related to the kits who showed up on the pond in June, but during the three months that they nursed, her nipples were visible. And if evidence of fatherhood were required, John, quite inadvertently, took a picture of the Inspector General's penis while helping me document a spate of scent-mound making. Finding myself unable to keep up with what was going on (all the animals were engaged in marking behavior at once), I handed John one of my cameras and asked him to shoot pictures of the Inspector General while I concentrated on Lily and the others. We both then clicked away as one animal after another shoved mud up on the shore and sprayed it, first with scented oil manufactured in its anal glands, and then with a squirt of castoreum produced by still another set of glands located in the beaver's cloaca. To excrete these oils, a beaver must first relax the sphincter muscle that encircles the opening to its cloaca. In so doing, the contents of this pocket momentarily protrude. In this instance, the Inspector General's penis became briefly visible, although John did not notice it at the time he fired the camera. Not until later, when I made prints from his negatives, did we gain irrefutable confirmation that the beaver I had referred to as a male from day one was indeed of the masculine gender. I must admit that I had a good laugh over John's unorthodox method of discovery. So much for taking blood samples, lining drugged beavers up in front of X-ray machines, and palpating the animals' cloacas.

Throughout February the beavers continued to live by their own clock. With every rotation of the earth, they woke up at a progressively later hour until their circadian rhythms of sleep and wakefulness were completely reversed and, for once, coincided with my own. This development sent me to a study on the winter activity patterns of two radio-marked beaver colonies in Massachusetts by Richard Lancia, Wendell Dodge, and Joseph Larson. These researchers found that some beavers, after being confined under ice cover for a period of time, fall out of sync with the twenty-four-hour day. One of their male subjects, for example, extended his activity/sleep pattern by five hours; a female extended hers by two. Thus the onset of that colony's waking life occurred at a progressively later hour with every revolution of the earth. During periods of thaw, however, when the beavers popped their heads out of ice cover and exposed themselves to a dose of daylight, they quickly reverted to their normal twelve-hour-wake, twelve-hour-sleep schedule and once again became nocturnal in their habits.

There must be survival value in the beaver's erratic winter biorhythms. Following a free-running clock, in fact, ought to allow a wintering colony to cut down on its food intake. For by pushing their waking time progressively forward—converting a twenty-four hour cycle into a twenty-nine hour one—a colony would experience only seventy-four activity periods during three months of winter, whereas were they to adhere to their normal twenty-four-hour schedule, activity periods during this same time frame would number ninety.

The question of how certain inadequately provisioned beaver colonies make it through severe winters has intrigued many scientists. Studies in the far north demonstrate that under worst-case conditions the species is capable of turning down its metabolic rheostat. One researcher monitored the body temperature of beavers in Alaska and noted a drop of 3.6 degrees Fahrenheit during winter.

Other adaptations also help the beaver live within a tight economy. No doubt being dressed in a warm winter coat helps the animal conserve body heat, and so reduces its need for calories, as does the fact that life in a one-room house, of necessity, gives rise to a good deal of huddling. Moreover, wintering beavers feed on their own body fat. My Lily Pond animals were noticeably less plump when they emerged from their lodge in spring than when they entered it in early winter. And finally, beavers have invented an efficient method for extracting the most possible nutrition from their high-fiber diet. It is called *coecotrophy*. Like cud-chewing animals, *Castor canadensis* eats everything twice. But whereas ruminants—deer, elk, moose, and caribou—do this by burping up partially-digested food and giving it another good chew, the beaver puts food all the way through his digestive tract before dining on it again. That is not to say that he ingests his stools. What he devours is a nutritious gelatinous substance, also passed from his anus. A beaver stool, being the end-product of double digestion, appears to be almost pure sawdust, whereas the porridgelike residue he excretes and re-eats is partly processed food that needs another go around.

While the beaver's practice of coecotrophy is not likely to endear him to the squeamish, it ought to be viewed for what it is—a survival strategy that allows the species to exist on a high-fiber, low-calorie diet of bark. And it certainly helps explain how certain beaver colonies manage to get through winter on short rations. What a beaver eats, a beaver absorbs!

Chapter Twelve

March came in like a lion, dropping three inches of wet snow on the park. Then, four days into the month, a cold wet drizzle coated the ground with a slippery glaze. Nevertheless, even though walking was hazardous, I was determined to visit the beaver house. But how to get there? I no longer trusted the pond ice to support my weight, nor could I make it to the lodge by land, for the dense laurel on the steep north bank was no more penetrable in winter than it had been in fall. In the end, I put my halogen lantern, my camera, my binoculars, and a few birch branches into my backpack, pulled on hip boots and then, hugging the shoreline, risked walking on the exceedingly thin and slippery ice.

The day was dark and a mist hung in the air. As I stepped onto the pond, I knew I would break through and my intuition proved correct. Twice I found myself knee deep in water. In the end, however, my effort was rewarded. For when I arrived at the lodge, I discovered a pool of water had opened beside it.

I tossed my token offering of birch branches on top of the weak ice that ringed this pool in the hope that a beaver would put in an appearance and climb out of the water to retrieve it. Thus I would have a rare opportunity to photograph the animal in a winter scene. Then I sat down on a wet rock beside the lodge and listened for beaver voices.

Ten weeks had passed since I had actually seen any member of my colony, and lately, since the colony had reverted to its same old nocturnal lifestyle, I had not heard much from them either. Had they wintered successfully? Had they stretched the two-week supply of food we brought them to last three months? Were all six beavers still alive inside the lodge?

While I sat quietly on a rock, wet snow began to fall. In such weather my camera is of little use, and so I packed it and my binoculars back into my bag and entertained myself by watching a pair of mallards waddle about on the ice. Migrant waterfowl were already returning. Then I waited in silence and allowed my mind to grow quiet. After a short while I heard a glug. A beaver had dived into the plunge hole that led into the pond and was about to make his debut in the open water directly in front of me.

The prospect of seeing one of my subjects again started my heart pounding, but I managed to hold still while I waited to discover who would surface. I was more than a little apprehensive that my presence so near to the lodge would alarm the animal. I need not have concerned myself over that possibility, however, for what surfaced was a mighty sleepy-looking creature. He floated about for a few seconds, then, after shaking his head like a dog, hauled himself onto a snow-crusted rock just six feet from where I sat. Who was this beaver? He looked too thin to be the Inspector General and too large to be either of the yearlings. And Lily I would recognize, fat or thin. Her soft gray muzzle and gentle questioning expression were characteristics that were uniquely hers.

The sleepy-looking beaver, draped on the rock, let his eyes close part way, and did not move at all. I, too, remained absolutely immobile for I don't know how long. Wet snowflakes, falling on my face and eyelashes, blurred my vision. Sooner or later I would have to raise my hand and wipe them away, a movement likely to startle the beaver. Better to alert him to my proximity with a quiet word or two in the hope he might hear something familiar and reassuring in my voice. And so I spoke.

"Don't be afraid," I said quietly. "You remember me."

And apparently he did, for after gazing at me for a few moments, he turned his sleepy attention from my wet face to his own, which he then proceeded to groom with his front paws. While I watched him wipe his cheeks and rub his ears, I failed to notice the appearance of another beaver. Silently, one of the kits had descended the plunge hole and come up for air beside the thin, dark adult. I kept right on talking.

"Well, hello there," I greeted him. "Glad to see you again, too."

The baby seemed as unperturbed by my presence as the adult. Both animals, in fact, appeared somewhat stupefied, as if dazed by the unaccustomed sensory load of light and odor and sound. Being icebound for three long months must have had a tranquilizing effect on them. As I continued to speak, they remained absolutely still.

Then, noiselessly, a second kit showed up and, as had the first, lined up beside the thin dark adult. I was delighted to know that both late-born youngsters had made it through winter despite the paucity of food in the family larder. And they looked to be in good health. The appearance of the second kit incited the first one to play, and in short order the two were rolling about in the water. I got the impression they were celebrating their new-found freedom and that they reveled in their own buoyancy and agility. And who wouldn't? Given another life, I might like to be a seal or otter or beaver—and for the same reason.

After a time, one of the kits swam over to me. By now he was fully alert and fixed his gaze directly on my face. Whatever he saw there apparently did not frighten him, for he quickly became distracted by the sight of the branches I had tossed onto the ice and went over to investigate them. He did not climb out of the water, however, as I had anticipated, but instead grasped the edge of the ice shelf with his front paws and pressed down on it, as if to break off a piece. When this did not work, he swam underneath the ice shelf until he was exactly below the spot where the branches rested. Then he rammed his head against it. After a couple of tries the ice shattered, and the little beaver popped up through broken shards and reached for a branch, which he then pulled underwater. A few moments later I heard gnawing sounds from inside the lodge.

So even baby beavers break ice! And how adept this youngster had been at it. Was the impulse to do so encoded in all beavers? Perhaps, but even so there was more to what I had seen than rote behavior, for there was nothing mindless about how the kit had approached the problem of getting the stick. Upon discovering it was beyond his grasp, he attempted at first to use his hands to dismantle the ice shelf upon which it rested. When this failed, he

Blossom studies me before attempting to retrieve the branches I have tossed onto the surrounding ice cap.

tried another tack. He swam underneath the icecap and punched it out from below.

No sooner had the little beaver brought evidence into the family's living quarters that birch branches were to be had, than the big, thin beaver, who meanwhile had returned to the lodge, emerged again. Like a navy icebreaker, he nosed his way into the floating shards and seized all the remaining branches. Clutching this twiggy bouquet in his teeth, he dived under the ice cover and vanished.

I fished my binoculars from my pack and scanned the shores for him. Falling snow made viewing difficult, but after a time I spotted him at my old station across the pond. He had swum eighty yards under the ice cap, emerged by way of an otter hole, and was peacefully eating his find where no kits could pester him for a share.

It was almost dark when I left, and as I made my way back along the south shore I once again broke through thin ice up to my knees. It was with a real sense of relief that I reached the dam. Upon climbing onto it, however, I saw that a band of water, nearly a foot wide, had opened alongside it during my brief absence. How had that happened? As I walked the slippery crest, that gap of water made me nervous, and I had to steal a stick from the back side of the beavers' engineering work to give me support. With snow blowing in my eyes, I did not have much of a sense that winter was on its way out; yet a number of signs were pointing to that fact. Though the cold season was putting on a brave front, warm currents were at work under the ice, eroding it from below. Soon those two mallards I had watched earlier would be nesting in the saw grass.

I walked in four inches of snow a quarter mile down a closed park-road to where I had parked and rejoiced all the way. At least three members of my beaver colony had survived the killer season. And ice-out was near at hand.

Chapter Thirteen

I cannot help but believe animals experience joy. It seems so obvious. When my dogs celebrate my return with high leaps, or bolt from my car and race madly about the yard of my weekend cabin, I am entitled to believe what my eyes and common sense tell me: they are expressing joy. Their behavior is so analogous to our own (jumping for joy or wildly running down a beach) that it would seem sheer sophistry to pretend they do not actually feel the emotion they so vividly portray.

And so it was with the beavers. One afternoon in March I arrived to find the pond ice had gone out. All of it. Not a patch or a shard was left floating on the water's surface; nor did any icy slabs lie grounded along the pond's banks. And the still blue water perfectly mirrored the sky as it had never done when the lilies were up. I was amazed to see such an expanse of pond, for I had only known it in full dress, an aquatic garden through which the beavers traveled on narrow swimming lanes. This condition would not last long, however. The fast-growing lily plants, which I could see sprouting underwater, were already three inches high and reaching for the light. In another month they would begin to dot the pond. In another two months they would blanket it. Meanwhile, the beavers enjoyed unimpeded movement, and for however long they would lack floral cover, I would be able to spot them with ease.

And so I did. The first beaver to emerge into the open water was not the Inspector General, not Lily, not the kits, but the yearling, Laurel, whom I had not seen in nearly four months.

Though her frame had grown larger and layers of fat had melted off her round body over winter, I had no difficulty recognizing her. Her fur had always reflected glints of red and now, for some reason, it had deepened into a shade of auburn. In the years that followed I would observe seasonal color changes in all the beavers I watched. Reddish tones became more intense during winter, when coats are prime; blond tones darkened.

The Inspector General was next to exit the lodge, and he and Laurel met in the water. Immediately the two began porpoising over one another's backs, while slowly making their way to the dam. They were in no hurry to get there and when at last they did, they didn't remain long. After making a perfunctory inspection of that work, they retraced their route, passing their lodge, and then proceeded all the way to the pond's marshy inlet. While covering that three-hundred-yard distance, they engaged in nonstop aquabatics, plunging under and over each other, swimming together and down. Suddenly, up again. First one, then the other, rolling, porpoising, somersaulting. This was exuberance. This was release. This was animal pleasure.   ·

Well, I thought, they ought to be happy. It's spring. Now at last they can breathe fresh air and see light. What's more, for a brief time they are not constrained to stay within swimming channels but can move about as they please. They can dive and surface and disport in four whole acres of water, like I've never seen them do before.

Next Lily came out of the lodge, accompanied by the yearling I called the Skipper. And the two of them behaved in the same manner as had the Inspector General and Laurel. Then I knew that all the beavers had wintered successfully: not one had been caught in a trap, not one had starved, not one had died of old age or disease or suffered an accident. And my own joy corresponded to that which the beavers were so gaily expressing. Spring had come.

While daylight lasted I walked along the banks and looked for peeled sticks, driftwood that would tell me what the colony had survived on under the ice. Much of this debris, however, would not have washed ashore, would still be lying in a midden heap at the bottom of the hole the beavers had dredged to contain their winter food stores. In time some tooth-etched sticks would be raised

Lily and the Skipper take in the outside world after a long winter's confinement under ice.

and used to patch worn places on their house and dam. Beavers are remarkable in this respect. They make use of waste products. Debarked food sticks become lumber. Wood chips, produced when trees are felled, are used for bedding. Mud, excavated during channel dredging, is packed onto their dam to stop leaks.

Judging from the small number of pared sticks I located along the shore, however, it looked as though the beavers were going to be short on handy lumber this spring. What few I found obviously were the remains of the aspen John and Dan and Nina and I had donated, for their cut ends were not gnawed, but had been neatly severed by saws. Meanwhile, I made a surprising discovery. The shoreline was littered with hundreds of blackened lily rhizomes, refuse that had washed up after ice-out. And each one of these had been partially eaten. So *that* was how the beavers had made it through winter! I examined dozens of the long fibrous roots, broke them apart and looked at their insides. They had the consistency of raw potatoes. Clearly, the beavers had dug these swamp roots from the bottom muck and been sustained by them throughout their long imprisonment. So our donation of aspen branches had not been necessary after all!

Obviously, there was a lot I did not know about beavers' food habits, and I went back to the literature to see what others had discovered regarding the animal's use of aquatic roots. I learned that southern beavers, unlike those in northern states, do not lay in winter food stores. That made sense. Why would a beaver waste energy caching food where pond water does not freeze and shore vegetation flourishes twelve months a year? Still, even in the South certain plants become dormant seasonally. Do these provide high-grade nutrition? Might beavers dig up and eat rhizomes during the dry season? I found no information about this.

Then I reviewed Françoise Patenaude's study of the Canadian beaver, a portion of which I had omitted, since her study had been published in French and was hard going for me. And there it was. While looking through a one-way window at her wintering colony in Quebec, she observed animals bringing "*rhizomes de nenuphars*," or the roots of the bullhead lily, into the lodge and eating them. That the plant she named was not the same aquatic lily that grew so profusely at my pond was of no consequence. The subjects of her study had dug rhizomes from "*le fond du plan d'eau*," the pond bottom. Beavers could do this, then. By some mysterious means they are able to locate edibles buried deep in bottom muck and dredge them up with their front paws. I would not have to worry about my colony ever again, for they would certainly survive as long as lilies grew in their pond. And judging from the super-abundance of *Nymphaea odorata* there, that would be a long, long time.

Over the next three weeks, the beavers did little else but make scent mounds. There was a practical point to their doing this. The mounds served as "no trespassing" signs, warning emigrating beavers that the property was occupied. And in April late-yearlings (animals that have overwintered twice with their parents and who are approaching their second birthday) depart their natal ponds and seek home sites of their own.

Most would travel alone, but some would venture forth with a sibling. Many would cross wide stretches of dry land and colonize new watersheds. Some would chance onto unoccupied sites on rivers and lakes, where natural water levels required no dam be built. Others would have to work hard to create a pond from a mere trickle of a stream. Still others might happen upon an intact, but no longer occupied, beaver pond; its former residents having been trapped out. Come winter, these seemingly lucky animals would in all likelihood meet the same fate as the site's previous tenants. And they would not be the only fortune seekers to fall victim to man. Inevitably, some enterprising animals would discover how, by plugging up irrigation ditches, a farmer's cultivated fields could be converted to a pond. Still others would stuff sticks into culverts and so transform roadways into streams. Such "nuisance beavers," as they are often called, bring out state animal-damage-control agents, who put an end to their endeavors and to them, as well. Moreover, a good many animals would find no unoccupied or suitable habitat, no matter how many miles of watershed they explored. And so, in the end, a large percentage of dispersing beavers would number as nature's casualties. And this is as it should be, for, like all species, beavers are endowed with biological and behavioral mechanisms to keep their populations in check. Don't ask, "What eats the beaver?" Lack of suitable habitat eats the beaver.

At the same time nature has more than one string to her bow. She has endowed the species with strong social glue, and a few unsuccessful emigrants find salvation in this. For whatever inner pressure motivates young beavers to strike out on their own in the first place does abate with time, and then the pull of family ties, or perhaps the tug of place, surfaces, and many unpropertied wanderers return home again. What is most amazing is that they are accepted by their parents, even after an absence of months or even years. That beavers can recognize their grown offspring after prolonged separation is impressive enough; that they are willing to share their limited resources with progeny of breeding age is mind boggling. Yet this behavior, first documented by Harry Edward Hodgdon, who studied marked beavers in the Quabbin area of Massachusetts over many years, has also been reported by a number of other researchers from the University of Massachusetts at Amherst. And in time I too would witness it.

Obviously then, beavers make sharp distinctions between "their own" and any alien beaver who encroaches on their borders. While accepting the former, they make every effort to dissuade the latter from making so much as a rest stop at their waterworks. They usually manage to banish the stranger without having to resort to aggressive action, simply by depositing their unique brand of scent repellent in conspicuous places about their property. Any wandering beaver who travels through such posted real estate will do so with dispatch. In fact, the pace at which such an animal departs is normally so rapid that the pond residents do not even bother to chase him.

The whole point of scent marking is to prevent bloodshed, for a species whose front incisors can cut through ironwood needs strategies by which to avoid using these terrible weapons against its kind. There is no survival value in intraspecies warfare, for those beavers who are drawn into battle often suffer dreadful injuries that prove fatal to both combatants. Scent marking is a peaceable means of signaling that a piece of watershed has been claimed, and this message is nearly always respected by itinerant beavers.

And so in April the Lily Pond beavers set about posting their waterworks. On prominent rocks or fallen logs, on grassy hummocks and knobby tree roots, they deposited blobs of mud or wads of leaf-and-stick detritus and then anointed these cairns with oils expressed from their anal and castor glands. The Inspector General was the most active marker. Sometimes he worked with haste, shoving one glob of bottom muck after another up onto the shore and rapidly waddling over and squirting each one with his spicy scent.

On other occasions, he created his scent mounds with more ceremony. Sometimes he would place one a few feet inside the shoreline and atop some flat stone or elevated tree root. This required him to walk uphill on his hind legs while balancing muck on his short forelegs. If his load got away from him en route, he would make every effort to scoop it up and then continue on to his destination. He never anointed the mucky debris where it fell, and would abandon the whole endeavor if unable to collect what was dropped. Mostly he made it to his target site and once there he waddled across the deposited muck so that his open cloaca made contact with it. And while anointing it with oils, he emitted a loud fartlike sound.

No one knows exactly what or how many kinds of messages may be communicated by scent mounds. It seems these mud pies not only mark property and deter aggression, but also serve as a means of attracting mates. Recent findings demonstrate that beavers are able to distinguish between male and female anal secretions, and studies done by the Amherst researchers has led to speculation that oils from the anal glands are as important as secretions from the castor glands in disseminating social information. It is thought that pheromones produced by the anal glands convey information about an individual's age, sex, and reproductive status, whereas secretion from the castor glands provides news about what the animal has been eating and, in addition, acts as a fixative to extend the life of the anal-gland secretions. (The perfume industry has long known about the fixative property of castoreum.)

But how do a beaver's scent mounds attract a mate, given their opposite function of warning away intruders?

An answer is suggested in a study by Harry Edward Hodgdon, which demonstrates that beavers can discriminate between scent markings left by males and those deposited by females. Thus, where one member of a breeding pair has died, a transient animal of the missing gender presumably could sniff out the news of the male or female "vacancy" and move in to fill it. Although beavers couple for life, neither sex wastes time acquiring a replacement for a dead mate.

That beavers react aggressively to the presence of a trespasser's scent mound was demonstrated in another experiment. D. Muller-Schwarze and fellow researchers introduced alien mounds into the territories of two beaver colonies with just that result. As soon as the resident animals got a whiff of the foreign odors, they began hissing and patrolling the shore where the strangers' mounds had been placed. After a while one of them summoned up courage to mount the bank and cancel out the unwanted scents with a blast of his or her own excretions. But that was not all they did. Sometimes they pawed at an experimental scent mound until it was obliterated. On other occasions, a beaver would salvage muck from an unwanted mound and incorporate it into one of his own making, which he then thoroughly anointed. Beavers are incurable recyclers.

The Inspector General reacts to an alien scent-mound on his home turf by destroying it.

110

I once saw the Inspector General behave like this and obtained a sequence of close-up photographs of the incident. I can only assume that he was responding to a foreign scent-mound left by a passing beaver when he mounted the bank in a state of high dudgeon, alternately hissing and emitting a gurgling sound I took to be his way of growling. With eyes wide and neck hair raised, he sniffed at the mound of mud, then proceeded to tear it apart with his front paws. Having done that, he picked up some of it in his mouth, carried it down the bank, and dropped it into the water. Then he swam away with his head raised, sniffing the shore for evidence of more bad news.

Even while I was enjoying the knowledge that all my beavers had wintered successfully and were now signaling their intention to remain at Lily Pond by busily marking its borders, Laurel and the Skipper disappeared. It happened all at once. One night the two of them were anointing scent mounds, and the next night they were gone. Unlike their parents and younger siblings, who seemed not to notice their disappearance, I missed them and walked the full length of the watershed in the hope I would find them somewhere along its course. But I didn't. Of course, it was time for them to leave. They were almost two years old.

Meanwhile, I comforted myself with the thought that replacements were probably on the way in the form of newborn kits and I created a route through the laurel to the beaver house. It was my intention to station myself nearby and listen to what went on in a beaver lodge while infant beavers were being confined there.

Chapter Fourteen

M any writers have unwittingly perpetuated a myth concerning the role of the father beaver during and after his kits' birth. Relying on what other misinformed writers have published, they report that he departs the lodge during parturition, takes up residence in a bank burrow, and does not rejoin his mate until the babies are several weeks old.

That is not at all what happened at Lily Pond. Nor does it conform with observations made by the Amherst biologists. And it certainly is not what Françoise Patenaude saw while peering into a beaver lodge in Quebec. On the contrary, all who have watched beavers closely have seen that the father beaver takes an active role, not only in the care of his infants, but in the actual birth process.

The late Dorothy Richards, in her book, *Beaversprite*, tells how Hunk, the male, remained with Chunk, his mate, throughout the time she delivered two kits. These animals, of course, were Dorothy's house beavers and might be expected to behave somewhat differently from those in the wild. But Françoise Patenaude, while peering into the lodge of wild beavers, reported exactly the same thing. What follows is a translation of her observations:

> Around three or four days before delivery, all members
> of the family collaborated to fix up the living chamber.
> Its walls were plastered, branches that protruded were
> gnawed, and the floor was covered with fresh grass. Birth
> took place during the day. The father beaver and one

yearling formed a triangle with the mother, presumably to protect the newborns. The other yearlings were allowed to remain in the lodge without taking part in the care of the infants. Soon after birth, the female licked each kit. In one colony, a yearling assisted in this task. [The only record of a beaver licking anything.] The female ate the placenta in several stages. She was the only family member to consume it (the only carnivorous activity seen). The newborn kits were miniature adults, completely formed, their bodies covered with fur, their eyes open, and some minutes after birth they were able to walk and swim.

My first clue that a litter of young had been born to Lily and the Inspector General was my inability to get a full count of the colony. Every night an animal was missing from my tally, although not always the same beaver. I took this to mean that one or another of the family members was remaining with newborn kits in the lodge. To obtain confirmation of my hypothesis, I pushed my way through laurel, seated myself on the south shore near the beaver house, and there was rewarded with audible proof that babies had indeed been born. Infantile mewling sounds were coming from inside.

Now I saw how active was the father beaver in providing food and bedding for his new offspring. Every evening he clipped a big bunch of the tall grass that grew along the north shore and, carrying this in his mouth, delivered it to the lodge. Was the grass he so assiduously gathered used only for bedding or was it also offered as food for the baby beavers? I read in one study that infant beavers begin to take solid food when only ten or eleven days old.

I longed to see inside the lodge, where the youngsters would be confined for perhaps three or four weeks, but I had to content myself with what my ears told me was happening in there. I pored over the scientific literature on the subject and learned that beaver babies come into the world weighing only about a pound, yet are fully furred and open-eyed, with their sharp little incisor teeth already erupted. Despite this advanced level of physical development, the kits are suckled by their mother for as long as two months

The Inspector General brings grass bedding to the lodge where his newborn kits are confined.

and may be kept in the lodge for nearly half that time, always under the watchful eye of an attendant.

There is something inconsistent in this. Normally an animal that is so well developed at birth does not require such close supervision from adults. The newly hatched chicks of certain ground birds, grouse and quail, for example, run about and peck seeds soon after cracking out of their shells and need not be kept in a nest and fed, as are the naked offspring of robins and bluebirds. Animals that are born ready-to-go are called precocial and, from all outward appearances, it would appear that baby beavers should fall into this category, for they can see, hear, walk, and even swim when only hours old, and after five days of life are sufficiently well coordinated to engage in wrestling matches. Yet they are confined to the lodge, sometimes for as long as a month, during which time every adult and yearling in the colony invests time and energy guarding and waiting on them. Why?

Françoise Patenaude makes an observation that sheds some light on this paradox. Conceding that beaver kits are indeed precocious at birth, she nevertheless states that the slow maturation of some movements and slow development of certain behaviors implies that these must be *learned* and she goes on to suggest that, during their long confinement in the lodge, the kits imitate their elders' manner of grooming, feeding, construction and even their reaction to predators.

It would seem, then, that the beaver, like a good many other highly intelligent creatures, is endowed with two sets of survival strategies—encoded behavior and the ability to discover and apply new responses to an old situation. Each strategy confers advantages as well as disadvantages on its possessors. Those precocial species that are highly programmed, although up on their feet and ready to go shortly after birth, lack flexibility. Should a shift in climate or some agricultural or industrial development reshape their habitats, such animals cannot adapt and are quickly extirpated. Conversely, animals that must learn to meet life's exigencies through trial-and-error or by imitating their caretakers are more vulnerable during infancy and, as adults, must invest more time and energy in rearing offspring than do precocial species. By the same token, however, they possess greater flexibility. Unburdened by a large repertoire of encoded responses, they can meet change with change and so are able to adapt to a wide spectrum of habitats.

Thus the versatile beaver colonizes a variety of watery niches from the Arctic Circle to the subtropics, and in so doing does not waste energy in mechanically performing activities that are irrelevant to the circumstances at hand. River-dwelling beavers, who have no need to create their own water levels, do not spend calories compulsively building dams; beavers in the southern states, where ice does not form on ponds, do not mindlessly lay in winter food caches; and my Lily Pond beavers, whose diet of lilies nourished them the year around, rarely bothered to cut trees.

Still a question remains: why, if baby beavers are learning animals, are they as well developed at birth as species programmed for early independence? Why are they born with their eyes open, their teeth erupted, and the ability to make paddling motions? Did *Castor canadensis* modify its survival strategy at some time in the past? Did it *add* learning to its encoded bag of tricks? If so, the question of whether the beaver works like a robot or solves problems like an intelligent being might be resolved.

To my mind, the beaver employs both strategies. It is certainly true that at times the animal behaves quite automatically. In an elegantly designed experiment, Swedish ethologist Lars Wilsson demonstrated that captive beavers with no previous building experience could be stimulated to build a dam in response to a sound recording of running water. On the other hand, I have no lack of evidence that my wild beavers, once they had impounded sufficient water to meet their needs, refrained from making further repairs on their dam regardless of how much noisy spill flowed over its crest. The species, it appears to me, can plug into two operational modes and for this reason it defies definition.

Most researchers today agree that the beaver is a learning animal and that the long investment adults make in their offspring is of critical importance to the species' survival. Two of the Amherst researchers, R.A. Lancia and H.E. Hodgdon, write: "The young express at an early age many adult behaviors. However, a long period of development in the family is required to perfect the behavioral repertoire, especially construction activities."

And so at some cost to themselves, beaver parents share food and habitat with offspring who are one and two and sometimes even three years old. In return, these hangers-on help in the care of the neonates; for every member of a colony, male and female alike, takes a turn at baby-sitting. In fact, judging from the traffic I watched entering and leaving the lodge, no sitter abandoned his or her post until a relief shift arrived. Sometimes that meant a tour of duty lasting several hours. Even Lotus and Blossom, who were still kits themselves when the next litter came along, assumed their share of this responsibility.

I wondered how attendants occupied themselves during the long hours spent inside the natal chamber. Once I saw a beaver push soiled bedding into the water, then some hours later return the "washed" grass to the lodge. I also noted that a relief shift would often arrive with a leafy branch in his or her mouth. Afterward the sound of gnawing emanating from the living chamber indicated that more beavers than one were dining on it; thus I presumed the kits had been brought food.

To gain more insight into lodge life, I again turned to Françoise Patenaude's study. She reported that the adult or yearling on duty always kept a close watch on the litter and physically retrieved any baby who fell into a plunge hole. Picking up a straying kit by mouth or nudging it to safety by nose were only two methods by which a sitter dealt with such an emergency. On occasion, a youngster was lifted up and carried on the forepaws of its rescuer, who then walked bipedally up to the dry chamber floor.*

Indeed the most important function of the baby sitter is that of life guard, for despite the fact that infant beavers are capable of swimming, they are unable to close their nostrils or ears, and even Dorothy Richards's protected house pair lost one of their kits to drowning. Moreover, the fur of neonates is not waterproof and any baby who gets wet must be groomed by an older beaver to prevent it from becoming chilled. Finally, the value of attendants becomes apparent when one examines the design of a lodge. As stated earlier, to exit this structure an animal must dive through a long plunge hole that opens at some depth below the surface of the pond, a maneuver infant beavers are too buoyant to perform. Thus they could not escape unassisted should water levels rise and flood their living quarters or should a predator gain entry to their natal chamber.** To save themselves, they would have to cling to the fur of an adult or yearling and be towed outside. That an adult would be inclined to perform such a service for a lodgebound youngster can be presumed from the fact that tired kits are allowed to piggyback their elders during early ventures about the pond.

Once outside, the buoyant baby beaver faces a less obvious, but just as challenging obstacle—how to get back in. Dorothy Richards, in *Beaversprite*, described a kit that was incapable of doing just that. As it happened, Dorothy had obtained the little beaver from his

---

*One characteristic that distinguishes the North American beaver, *Castor canadensis*, from the European beaver, *Castor fiber*, is its greater propensity to walk upright and carry objects on top of its forelegs. The European beaver and the North American beaver are actually two separate species, exhibiting morphological and chromosomal differences. The genus *Castor* is thought to have originated in Europe ten to twelve million years ago. During the Pliocene, some members of this genus crossed a connecting land bridge to North America and evolved into *Castor canadensis*.

**It is not known if otters attack beaver kits; in any case, during its long evolutionary history, *Castor canadensis* probably has been preyed upon by other aquatic animals. Certainly, some land predators are capable of breaking into a beaver lodge. I watched a bear try to do this in Alaska.

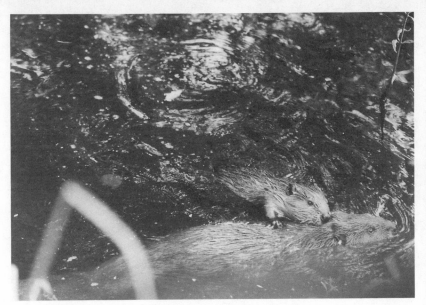

An adult beaver tows a tired kit.

wild mother when he was only days old. For seven weeks she cared for him, at which time she brought him back to his natal pond, where she hoped he would be accepted by his family. For several hours she waited onshore until every member of the colony had emerged from the lodge and come over to inspect the little home-comer. A touching reunion between mother and kit reassured Dorothy that her charge would not be rejected, and when his litter-mates (who were considerably larger than he) began to porpoise about with him, she sneaked away.

But when she visited the pond the following morning she found the kit all alone outside the lodge and mewling for help. Upon seeing her, the forlorn little creature raised his short forearms to be lifted up and carried back to his old rag bed in the kitchen (an orphaned beaver kit once melted my heart with this very gesture), and of course Dorothy obliged.

It now occurred to her what the problem was. It was not that the youngster had been denied entry into the lodge by the others; he simply wasn't able to make his way inside. To do so required he dive to a depth of three feet and remain submerged long enough to locate the entryway. But having been reared in a bathtub, he had acquired no such skill. And when the others vacated the pond for a day of sleep, he had been left behind. The solution to his problem, of course, lay in supervised practice, which Dorothy provided, and in due course, the kit was returned to the wild.

All of this information, in conjunction with the sounds I was hearing, enabled me to gain some idea of what was going on inside my colony's living quarters and why. By ear, I guessed (correctly) that two youngsters had been born to Lily, for the whines of infants are easily distinguishable from the vocalizations made by older beavers, and when the newcomers chorused I heard two voices, but never more. I noted that whenever Lily returned these contact sounds grew most intense, then stopped entirely, and I pictured the babies nestled against her body, quietly suckling. I grew impatient for the youngsters to make their appearance. Hope Buyukmihci wrote me that she had seen kits as young as two weeks old out in the water. Other observers have reported longer confinements, and I took these differences to be yet more evidence of the species' versatility.

About this time, the Skipper and Laurel came home again. I had not seen them in weeks, and when I first sighted them munching lilies on the pond, I was not convinced that they were the missing two-year-olds. But their colors and size exactly matched those of the émigrés, and when the Inspector General and Lily readily admitted them to the lodge where the new kits were still being confined, I had no doubt as to their identity. They assumed babysitting duty just as if they had never left home, thus granting other members of the family more relief time.

I have no way of knowing why those two-year-olds returned, but I suspect the existence of new kits played a part in luring them back. Beavers are extremely responsive to infant kits, and their solicitous behavior does not begin to wane until the young are six or seven weeks old. Until then the youngsters are immoderately indulged.

This time I was on hand for the babies' coming out party, though it occurred in dim light and I had to watch it from across the pond. Every member of the family was present to accompany the two youngsters up and down the donut ring. The little ones swam like corks and frequently clambered onto handy backs. To my delight, over the next two weeks, this performance was repeated many

Lily keeps a close eye on her infant kit during its first excursion out of the lodge.

A ditched youngster, upon finding himself alone, swims directly back to the lodge.

120

times. Whenever the kits emerged from the lodge, beavers converged from all directions and seemed to compete with one another for the privilege of shepherding them about the pond.

While the new kits seemed particularly attached to Lotus and Blossom and frequently followed these two older kits around, Lily was their best teacher. Normally, she and the babies emerged from the lodge together and immediately began porpoising in front of it. This exercise no doubt strengthened diving muscles and the kits appeared to enjoy it greatly. Though Lily did her best to keep both babies on the shoreside of herself, they were irrepressible creatures and would dive under her belly and climb onto her back or her tail, hitch rides, and generally do what they pleased.

In the beginning, while her two offspring were still infantile and inexperienced, Lily escorted them back to the lodge before going off by herself. After two weeks of touring the donut ring with them, however, she became less protective and even began ditching them in odd places. Though this alarmed me, I soon saw that Lily was doing no disservice to her young. By that time they knew their way about the pond and usually, upon finding themselves unchaperoned, headed straight for home.

Chapter Fifteen

One day a fisherman came to Lily Pond, and the Inspector General became entangled in his line. Until that happened I had kept out of sight of any visitor to the place, since I didn't want to call attention to my project or to the beavers. Also, I used such opportunities to spy on my colony and see how they responded to human beings who were not associated with me. For some time I had worried that my recurrent presence along the bank might have the effect of blunting their natural fear of *Homo sapiens*.

My suspicion proved on the mark. On this occasion, the Inspector General's immediate response to the fisherman's presence was to assume the beaver-alert posture, head elevated and nostrils working; but then he swam over to the shore where the man was casting his line and, after giving him the once-over, did not so much as slap the water with his tail. Instead he began munching on a lily pad, not thirty feet from where the stranger stood.

By now lilies had begun to appear on the surface of the pond, but only in patches, so the Inspector could not avail himself of vegetative cover and was as easy to spot as a crow on snow. Through my binoculars I saw that the fisherman was well aware of him—was eyeing him, in fact.

This is my fault, I told myself. These animals have become so habituated to me that they are now indifferent to the presence of others of my kind. And though I was fully cognizant of the fact that in this protected park it was against the law to carry firearms or set traps, I nevertheless feared for the safety of my colony.

For a time I simply watched as the man continued to cast and the Inspector continued to drift about in close proximity to him. Then quite suddenly, while pondering what might be done to disabuse the beavers of their innocent, misplaced trust in mankind, I noticed that the Inspector General was in trouble. He had swum into the fishing line and become entangled in it.

Immediately, I made my presence known to the fisherman, who appeared to know how to deal with the situation. He let his line go slack, "allowing the big beaver plenty of play," in his words, and somehow the Inspector General managed to get free.

"He was trying to take my fish," he afterward stated in an effort to explain what had happened.

I contradicted him.

"He wasn't after your fish," I said. "Beavers are vegetarians. It's just that this is his pond. He made it by putting up that dam over there. And right now he's using this particular part of it because water lilies, which *do* happen to be a part of his diet, have come up here first. What is wrong about all this is that he isn't afraid of you, as he should be, and I'm afraid that's my fault."

It was clear that the man was not convinced by my words, but he was a sympathetic individual, not a person who would deliberately harm a beaver. Nevertheless, the incident pointed up how the activities of even benign people can sometimes result in unforeseen disaster for wild creatures that come into contact with them.

That weekend John came down from Massachusetts, and I apprised him of what had happened. Then I delivered a kind of prepared speech, informing him of a course of action I had resolved to take.

"It's been wonderful being accepted by the animals and watching them go about their business as if they weren't being observed. I've seen more than I would have believed possible, all because the beavers have tolerated me so well. But now, since fishing season opened, a lot of people have been coming to the pond, and the beavers are not at all shy of them. I'm worried about what might happen when the news gets around that there is an active beaver colony here. I'm afraid the time has come for me to undermine the beavers' confidence in me, even though doing so will make future viewing of them exceedingly difficult."

"How do you propose to do that?" John asked with a look of astonishment on his face.

"I think I can do it by communicating with them in a language they understand."

That evening John and I set out for the pond, carrying a canoe paddle and a shovel. Our plan was to let these two pieces of equipment serve as our tails and use them to slap the water. I felt terrible about what we were going to do. The beavers trusted me, and I needed their confidence in order to watch them close-up. In a perfect world, the relationship that had developed between the colony and myself would be proper. But the world is not perfect. The relationship man has developed with animals is unnatural, out of balance, heavily weighted in man's favor.

Less treacherous predators than ourselves can and do live in close proximity with their prey. Wolves, for example, follow the caribou herds they feed on and, except when aroused to hunt, evoke no alarm in their victims, for the caribou can trust the wolf not to decimate its numbers. When a species relaxes its guard against human beings, however, it puts its ultimate survival in jeopardy, for modern man behaves erratically, unpredictably, and greedily. He no longer meets animals on equal ground. Having extended his sight and reach and grasp with the help of lenses and wheels and bullets, he is capable of destroying far in excess of what any wild predator could run down and kill. It is incumbent on man, therefore, to curb his predatory impulses, to be accountable to nature, to exercise the kind of restraint any superpower owes those who have been outstripped.

A great many human beings, however, lack insight into how distorted our relationship with nature has become and perceive wild animals as existing solely for their own pleasure and use. They defy what protective regulations and authority are in place and feel no qualms about poaching animals in our national and state parks. These thoughts and considerations were what now motivated me to undo my special relationship with the Lily Pond beavers—to warn them of man's treachery.

John was the first to create a fountain-high splash with the canoe paddle. He stood on a rock in the donut ring and waited until Lily approached and was about to cruise past him. Then he smacked the wooden paddle on the pond's surface with such force that it split down the middle. Even before the spray of droplets it raised had fallen back to their source, Lily dipped out of sight and presumably sped off underwater.

"So much for *your* tail," I said, readying myself to perform the same act with the more durable steel shovel. I had my eye on the Inspector General, who by now had completed his nightly check of the dam and was moving in our direction along the north shore. When he was thirty feet from me, I raised the shovel above my head and brought it down with all my might. Water splattered in every direction, and the Inspector General nosed to the bottom. By the time the big beaver surfaced, John had seized my shovel and was getting ready to execute another mighty thwack. His movement, however, caused the Inspector to let fly with his own tail, which, being better designed for the purpose than was our garden tool, sent a geyser of spray high into the air. Then he, too, vanished from the scene.

"I suppose we'll have to repeat this a few times to drive our point home," John said, as he took another whack at the water.

"Yes, but don't slap your tail until a beaver comes near us. Otherwise they'll get used to the sound and ignore it. Besides, we want them to associate this signal with us."

I need not have bothered to offer that advice. The noise seemed to have attracted, rather than repelled, Blossom and Lotus, both of whom came steaming toward us from the far shore. The spectacle caused John to straighten up and ask, "Are you sure this means what you think it means? It looks to me like we just put out a 'Calling All Beavers' announcement."

I was dumbfounded to see the sibling pair slow to an idle some twenty feet in front of us and then eye us with ingenuous curiosity.

"Do it again," I prompted, but now some doubt had crept into my mind that our scheme would have the anticipated effect.

John raised the shovel and, even as its scoop struck the water, Blossom and Lotus slapped their tails so that three sprays rose simultaneously in the air. Following this demonstration of their keen sense of precision, the siblings dived, remained underwater for a few seconds, then surfaced and waited for John to make the next move, which he did. Again he slapped the water and again the two young beavers performed exactly as before. Their tails hit the water at the precise instant as did the shovel.

"I think it's a game," I commented. "They're playing with us."

This was as interesting as anything I had seen beavers do. The youngsters seemed to be interacting with us—two members of another species. I recalled Hope Buyukmihci describing how she had watched baby beavers play with baby otters, and my surprise at her account. Now I could find no better explanation for what was going on than that these late-born beavers (technically still kits, for they would not be a year old until early September) were engaged in some kind of interspecies sport with us.

Suddenly I spotted two wakes on the open water.

"Here come the parents," I informed John, who stood poised to wield the shovel yet another time.

"It's like I said," he responded. "The message we're sending is, 'Calling All Beavers!'" Splat! He brought down the shovel.

I could not detect even an instantaneous lapse between the time John's "tail" and those belonging to Blossom and Lotus struck water. What close attention the beaver must pay to our most subtle movements, I marveled.

By then the Inspector General and Lily had arrived and appeared to both of us to be trying to corral their offspring and herd them into deeper water. Clearly, the two adults did not find the interaction between ourselves and their young entertaining. Both hissed at us.

"Well, at least we've turned *them* against us," I said.

But I was more than a little surprised at what I was seeing and hearing. For one thing, the Inspector General and Lily had always been mild mannered. Neither had ever hissed at me before. Nor had I ever seen them take action to protect their kits. In fact, it had always seemed to me that young beavers learn to distinguish and avoid threatening situations simply by being alert and imitating the reactions of their parents and older siblings. Whenever Lily or the Inspector General would make a dash for water, so did their offspring. Neither parent ever remained behind to nudge a kit to safety. Yet now they both seemed intent on doing just that. Of particular interest was the fact that the youngsters who had aroused their concern, although young for their age class, were hardly babies anymore. They had been dispossessed of that favored status in the family by the arrival of a new litter. Thus I was doubly surprised to see the parents approach a dangerous situation as if to rescue them.

"I'm giving this up," John said. "We've annoyed the adults enough so they won't be hanging around the likes of us anymore. And there's no scaring the young ones. They like this game."

My hope was that Lotus and Blossom, despite their obvious fascination with our slapping behavior, would come to adopt their parents' attitude toward us. Nevertheless, I decided to try one more trick to drive our message home. The next evening John and I set out for the pond with my two dogs on leashes. According to Dorothy Richards, beavers do not like dogs, having over long ages been preyed upon by a related species, the wolf. Now I wanted the Lily Pond beavers to see and smell and hear what kind of creatures we pal around with, in the hope that this would confirm whatever suspicions we may already have aroused in them.

I relied on Smiley, my small dog, but a good barker, to do the job for us. Zoe, my shepherd, I brought along for whatever doggy scent she would give off, for she failed to make distinctions between herself and other species. She early demonstrated this egalitarian outlook while I was on an extended field trip and she was being cared for by a family that counted among its members fifteen horses, twenty cats, four dogs, two ferrets, and a parrot. Not only did she make friends with all of them, she helped one of the cats tend a newborn litter of kittens. I knew Zoe would not bark at a beaver.

But on this evening, even Smiley's predatory instincts could not be aroused; he showed not the least interest in animals that spent their time swimming about in water. When we wanted him to look pondward, he stubbornly fixed his eyes on Zoe.

Nevertheless, our scheme was effective. The Inspector General was acutely aware that two dogs were by our sides as he patrolled the shore in front of us, hissing and occasionally slapping the water. Thus, once again I observed an adult beaver *approach* and try to disperse a threat, rather than make a quick escape from it. Meanwhile, no other beaver came near us during the three hours we spent at the pond. Whether this was because of some warning signal he communicated to the colony or an indication that our shovel act of the night before had been more effective than we realized, I cannot say.

In any case, after that all the beavers were shy of me. Moreover, lilies grew up and matted the water, concealing the animals from all but the most perceptive viewer. From my hidden cove I studied the people who came to the pond and noted how quickly after their arrival they became so engrossed in one another's company that they paid little attention to their surroundings. When three young otters captured and then noisily crunched up a painted turtle, not a single member of any of the parties that were scattered along the shore looked about to see what was causing the unusual noise.

Why do they come here? I wondered. But of course the beauty of the place provided an attractive setting in which to socialize. Moreover, some visitors to the pond enjoyed fishing for pickerel on the open strips of water the beavers had channeled. Some, however, came to do drugs, or to camp all night (illegally). On the Fourth of July a rowdy bunch spotted the beavers and threw firecrackers at them. Then I knew that my decision to discredit myself in the eyes of the colony had been a necessary sacrifice.

I did not intervene as the cherry bombs went off, their acrid odor tainting the lily-scented air. If my colony retained any ambivalent feelings regarding the nature of *Homo sapiens*, I expected those louts would, once and for all, dispel them.

Chapter Sixteen

I f Laurel and the Skipper had not returned home to see the
new kits, I might not have found the beautiful waterworks
they were creating; for as it turned out, they did not actually
take up residency at Lily Pond, but came and left erratically. One
evening I caught sight of the Skipper making his departure and
afterward I followed the drag marks his heavy tail created. This
trail led up the north bank, crossed a busy paved road, and then
continued down a steep embankment, at the bottom of which I
discovered a dammed-up stream and a pond in the making.

So the two-year-olds had been only a short distance from their
birthplace all along. Yet I had failed to find them. In trying, I had
made a common mistake. Operating from a wrong assumption that
emigrating beavers always follow the course of their own watershed,
I had searched up and down the small stream that fed into and
drained out of Lily Pond. But the two, in departing their natal
waters, had taken off across dry land and settled in a narrow valley
intersected by a small brook.

I called the place New Pond, and I was enchanted by it. Though
still very shallow and swamplike, it was rapidly being colonized
by representatives from all levels of the food pyramid. Insects,
hatching on the surface of the still water, were providing food for
a variety of birds and had also attracted aquatic as well as tree frogs
to the place. The noisy mix of bird and frog voices created an
interesting cacophony, not unlike that of a Chinese orchestra tuning
up. Through the din I could pick out the "wichety, wichety" of

129

the yellowthroat. Above my head kingbirds swooped after flying insects and shrieked insults at some cedar waxwings perched on a nearby tree. The little flock of waxwings looked as smartly military as cadets in their pointed caps and decorated with red-and-white wing bars on their khaki sleeves. Two wood ducks behaved like windup toys as they swam in and out and around some grassy hummocks that had not yet become immersed in the rising water. And the presence of a pied-billed grebe told me that this new pond was already deep enough in places to support a few fish.

On a slope immediately behind this new wellspring of life huge oaks and a large stand of pines dominated the landscape. These old trees were the end of a long succession of plant life, for they had grown so tall they had beat out the competition in whose under-stories they had sprouted. Now they stood shoulder to shoulder, creating so much shade and needle drop that no plant or shrub or sapling could gain a roothold here. Their glory days were over; still their long reign could not be overthrown by a new array of life. Now only fire or flood, lightning or old age could destroy them and so set the stage for the biotic cycle to begin again.

By contrast, the bowl in which the Skipper and Laurel were working had been enriched time and time again as a result of the water that *Castor canadensis* had repeatedly impounded in it; for this was not the first time the site had been colonized. Old stumps and beaver-scarred boles certified that the animal had been here twenty years earlier. And before that. The very contours of the valley revealed the impact that standing water had had in shaping the place over centuries of time. As a result, the soil here was exceed-ingly fertile; for each time that the trickling stream had been transformed by beavers into a quiet pond, its heavy load of seaward-moving topsoil had been trapped and had settled to the bottom. And during those beaverless intervals when the pond waters re-ceded, that rich flat bottom was uncovered and gave rise to a wide array of plants.

I studied the site that the Skipper and Laurel were now home-steading. Highbush blueberry, willow and alder, yellow iris, and jewelweed brightened the place. What a difference water makes, I marveled. What a difference the beaver makes, I thought to myself, as I watched the pair of two-year-olds dredge an underwater channel in the middle of their new pond.

So much needed to be done to make the place habitable before winter. Again and again the dam would be raised, whenever the water it impounded mounted and overflowed its crest. Not until sufficient depth had been achieved to cover the entryways to a lodge (which had yet to be built) would the beavers grow lax and fail to respond to the alarming sound of water spilling over their engineering work. Even before then, they would begin construction on their living quarters. To plaster this stick-and-mud domicile, they would use nearby bottom muck and thus excavate a handy basin in which to store the winter food they would next have to cut. Simultaneously, they would continue dredging channels. For without deep travel lanes they could not tow the branches they harvested to their food cache.

The two beavers showed no outward sign of experiencing the kind of anxiety a human being would feel if faced with such a work agenda. Few species, in fact, appear so oblivious to stress as does *Castor canadensis*. House wrens, for example, build their nests in a kind of frenzy, as if tyrannized by their seasonal timetable. Not beavers. The Skipper and Laurel proceeded at a leisurely pace. One handful of mud at a time, they scooped from the bottom of the pond. And, pressing this against their chests, they paddled slowly to the dam and shoved it up onto the crest. As unhassled as they appeared, however, they were in fact accomplishing two tasks at once—deepening a channel and raising the height of a dam.

Beavers work like that. Interrupting one operation to transport its byproduct to a site where that debris is wanted does not appear to confuse the animal, for the beaver does not become obsessed by the task at hand, nor does he look for the reward we human beings so enjoy—that of seeing a big job through to completion. Whether transporting heavy lumber, cutting trees, dredging, damming, or plastering, a beaver works for a limited time only before taking a break, whereupon a fellow colony member will, in all likelihood, show up and take over where he left off. Sometimes, after giving his all to a particular task, a beaver will quite suddenly lose interest in it and attend to an entirely different job. Nevertheless, he remembers the unfinished project and will even return to it *with the necessary materials in tow*. All waste products are recycled: dredged mud becomes house insulation or dam sealant; debarked food sticks become house or dam lumber; wood chips (fallout from a tree-felling operation) are brought to the lodge and spread on the floor for bedding. In this admirably relaxed manner, the efficient beaver accomplishes an enormous amount of work. Watching him is like attending a morality play, and I often thought I ought to take a lesson from it.

Now I divided my time between Lily Pond and New Pond, an arrangement that often left me wondering what I was missing at the other place. Lily and the Inspector General had begun damming the stream that fed into their pond, and as a result had flooded what I regarded to be a most unsuitable site. I did not see how they could possibly make a pond there, for by plugging up the channel that guided water into Lily Pond they had created a sprawling flow that moved in two directions at once: westward into Lily Pond and southward across land that sloped gently downhill. This situation required that they build two dams simultaneously in order to shore up a pond. To my astonishment, the beavers did just that. Both dams were completed in short order and abutted one another at a perfect right angle. I named this place Square Pond, for its north side was an old stone wall and to the east it was contained by a straight-edged bank. This adjunct pond was used for feeding, as it offered the beavers safe access to edible shrubs and saplings growing along its shores and provided them some relief from their steady diet of lilies.

A great deal has been written and some theories have been propounded about how and why beavers build dams. The Swedish ethologist Lars Wilsson theorizes that the sound of running water automatically releases stereotyped building behavior in the beaver. Using captive subjects, he demonstrated this to be the case by playing recordings of rushing water, which did indeed stimulate his subjects to construct dams in their pens and tanks. But not everyone accepts this explanation as the entire story. The French ethologist P. Bernard Richard, while in agreement that the sound of flowing water can act as a principal stimulus in releasing building behavior, nevertheless believes that these automatic responses are relayed by a higher psychic mechanism. In summary, Richard insists that the beaver's innate motor responses are put to use by the animal *to obtain a result*—an increase in water level. Were it not so, he says, they would lead to serious errors of adaptation. He draws on Wilsson's own study to support his view, citing an instance when Wilsson's experimental animals halted all work at stopping up a noisy *inflow* when their damming behavior failed to cause water levels to rise. To drive his point home, Richard quotes Wilsson's own explanation of that incident: "The animals seemed to learn to not react to the rushing water at the inlet of the brook since they did not get any positive result when working there."

I would have great difficulty were I to try to explain my animals' damming behavior as unqualified stimulus-response automatism; for if their motive in building dams was simply to silence the objectionable sound of flowing water, they certainly were not very consistent about doing so. For weeks on end they allowed their engineering work to fall into disrepair, and during these times they remained oblivious to the noisy leaks that developed. On one occasion I counted thirteen such breaks, none of which were mended over a three-week period. Yet every member of the colony swam past the structure every night. Then quite suddenly, just as I was beginning to despair that the Lily Pond beavers retained any of the impulses proper to their kind, they went to work and stopped up all the leaks in a single night. And the next night they added two inches to the crest of their already silenced dam. I could identify only one variable to account for the beavers' behavior: when the pond was at or above a certain level, they did not bother to work on the structure, no matter how like a waterfall it sounded; when water fell below a critical level, however, the colony promptly plugged up every noisy outpouring.

I have no doubt that the sound of running water can and frequently does elicit building behavior; Wilsson's experiments clearly demonstrate that. My observations suggest, however, that the beaver does not respond to this stimulus willy-nilly, but learns how and where and when to give rein to his building impulses so as to achieve a significant result.* For just as every wild kitten is born with a "pounce response," which is triggered by the sight of a scurrying object, that innate motor behavior has no survival value

---

*Were this not so, the animal might indeed be caught in the grip of an automatism that would lead to "serious errors of adaptation," as Richard puts it. Picture a beaver damming himself into his lodge in response to the sound of high waves sloshing into his entryways.

until its possessor learns to direct it toward a significant result (namely obtaining dinner). And since wild kittens do not automatically know how to hunt and kill, it is of critical importance that they learn from their mothers when and where and how to put their innate pounce-and-chase responses to practical use. That explains why young bobcats follow their mother about until they are nine months old and nearly as big as she is. Since young beavers stay at home for more than twice that time, it seems logical to suppose that they, too, acquire know-how from their parents and older siblings. If nothing else, the youngsters may learn when *not* to plug up a noisy gurgle, a sound that abounds in their sloshy environment.

Building a dam is not an easy thing to do. I say this from firsthand experience. One afternoon, while exploring New Pond, John and I decided to look for burrows on the far bank. To get there, we took a short cut across the dam that Laurel and the Skipper had built. But the fresh mud on its crest had not settled, nor had the structure been in place long enough to give rise to rooted plants that would hold it together, and so it failed to support our weight. When we were half-way across, a section collapsed. I was not so much upset over the dousing we got—which I felt we deserved— as I was over the damage we had done to the beavers' new dam.

"We ought to try to fix this thing before the Skipper and Laurel wake up," I suggested to John.

And so, after making a trip to my cabin for hoes, we pulled on our hip boots and waded out to the break we had made. In fairness to John, I must say that he did not think much of my idea, being convinced that the beavers would take care of the big break all by themselves. I knew he was right, but an urge to slosh about in a beaver pond had been growing in me (being a spectator all the time can pall), and repairing the dam we had damaged seemed just the excuse to satisfy it. And so we waded in with our hoes and tried to bring up bottom mud.

It didn't take long to discover how difficult a task we had set for ourselves. The pond bottom was rock hard, the ground having been inundated for too short a time to have softened.

"They must be getting their mud from somewhere other than here by the dam," John finally said. And so we began testing various places for muckiness.

"I found a patch," I called, after considerable hoe-pounding.

John joined me in trying to raise my find to the surface, but the stuff was unmanageable. It dissolved, and clouded the water as we tried to get hold of it.

"What about sticks?" I asked. "Maybe we should just try to push some sticks into the breach."

That idea proved no more workable. Water rushed over and under our haphazard placement of forked and branching lumber. In the end, we gave up trying to repair the dam and waited to see how the beavers would do it.

Their efficiency in accomplishing this feat was downright embarrassing. Within minutes of waking, the Skipper and Laurel were hard at work removing our poorly placed sticks and creating a more satisfactory arrangement of them. A long straight one was maneuvered to lie lengthwise across the breach. Then both animals scoured the shore for buttressing material. Several long polelike branches were towed to the break and heaved over it, so that they came to rest in an upright position on the backside of the dam. To secure

The Skipper shows John and me how to repair a broken dam.

these in place, the beavers held and guided them with their front hands, while pile-driving them with their mouths.

Afterward the Skipper packed finer material—branching twigs and bottom debris—against this picket work, while Laurel made repeated trips for bottom mud, which she obtained from a place John and I never would have found. Apparently, the two had dredged a tunnel under a small island, land that had not yet become submerged in the rising pond. Into this waterfilled tunnel Laurel repeatedly dived, and each time she surfaced she came up with an armload of fresh mud pressed tightly to her chest. The walls of the deep tunnel, not being in the grip of roots, had quickly softened, and so it was from here that the beavers mined all the mud they needed to cement and seal their dam.

I noticed that the order in which materials were used to fix a breached dam differed significantly from that used by the Lily Pond beavers in creating Square Pond from scratch. To stop fast-moving water, the Skipper and Laurel began with heavy lumber and only later used mud to fill in chinks and seal their repair. By contrast, the Lily Pond beavers, in containing a shallow, sluggish flood, first pushed up mud to form a long ridge. They then added fine twigs and decaying leaves to this. Not until several days had passed and the water they raised threatened to erode the soft structure did they back it with upright sticks.

In applying two different strategies to stem two different flow rates, the beavers seemed to have invoked some choice. Perhaps beaver kits serve long apprenticeships to acquaint them with the many ways that water behaves and to allow them time and freedom to experiment with different methods of dealing with what looks to us to be simple flow. To understand precisely what I mean, you would do well to pull on a pair of hip boots and try to mend a beaver dam.

Two damming strategies are applied to two different flow intensities. (1.) To stop fast-flowing water, beavers begin with heavy lumber and only afterward plaster this infrastructure with mud. (2.) To contain shallow, slow-moving water, however, beavers begin by pushing up a ridge of mud, which at a later time they back with sticks.

Chapter Seventeen

While all of this building was going on I got to know Lily's new kits, whom I named Huckleberry and Buttercup. Buttercup was blond and the cutest baby beaver I have ever seen. Her face was not so narrow and mouselike as those of some kits. Her fat cheeks, in fact, looked as if she had stored a few nuts in them, chipmunk-fashion. Her sibling, Huckleberry, was dark like the Inspector General and as he grew older he came to look more and more like him.

Lily always did me the favor of producing litters of two, each kit a different color. Laurel and the Skipper, whom I had met as yearlings, were auburn and a shade of medium-brown; Lotus and Blossom were blond and mahogany, respectively; and now Buttercup and Huckleberry had copied the colors of their next older siblings. This made it possible for me to recognize every member of my colony without having to mark them; for after determining an animal's age by its size, I could differentiate it from its same-age-class sibling by color. While the kits were very young, however, these color distinctions did not exist, their first soft coats being uniformly brown. Not until they had been out of the lodge a few weeks did their second coats grow in blond or auburn or medium brown or mahogany.

It goes without saying that I never knew the sexes of the baby beavers; nevertheless, I gave them male and female names and referred to each as a "he" or a "she." This helped me better remember individual animals. (Besides, I object to the literary practice of referring to an actual living animal as an "it," which in my view

Buttercup is blond and the cutest baby born during my four-year study.

relegates that individual to the level of an inanimate object. I do use "it" when describing species' behavior, however.) As it happens, beavers produce roughly equal numbers of male and female young, so my chance of guessing any individual's sex correctly was about even.

Throughout May and June, Laurel and the Skipper continued to divide their time between New Pond and Lily Pond, crossing back and forth so frequently that the grass lay flat on the paths they created. Were these siblings now to be mates? And if so, would this brother-sister union produce inferior, perhaps even deformed offspring?

It is a common practice for highly social species (the beaver being but one) to breed with a parent or sibling—a survival strategy by which isolated colonies produce offspring exquisitely adapted for life in their particular locales. And even though this practice results in genetic uniformity *within* each family group, it actually promotes genetic divergence from one inbred colony to the next, thus guaranteeing a healthy gene pool across the species' megapopulation.*

*In animals that do not inbreed, genetic variation shows up in *individual* animals across the species' entire range, whereas in animals that do inbreed it is expressed in *clusters* of animals.

There is little danger, therefore, that such species will, in times of pandemic disease or abrupt environmental change, become extinct for want of genetic diversity. In fact, faced with such a calamity, inbreeders may even have an advantage over outbreeders, for each cluster of like-animals serves as a reservoir of proven genotypes that will be faithfully reproduced.

But what of genetic depression *within* these homozygous colonies? After too much inbreeding, do not malformed offspring begin to show up?

Most geneticists agree that deleterious genes have little chance of becoming fixed in inbred colonies, for in wild conditions maladapted animals simply do not survive long enough to reproduce major weaknesses. Moreover, outsiders are from time to time admitted into these closed societies—when either member of a breeding pair dies and leaves no same-sex offspring to replace himself or herself, for example. Thus every colony receives an occasional infusion of new genetic material, but not so much as to swamp those valuable traits that have been selected for and preserved through inbreeding.

And so I was not at all concerned over the fact that Laurel and the Skipper had set up housekeeping together. Likely they would produce fit offspring, well equipped for life in a pond so near their birthplace. But as matters turned out, this union was not to be.

One day I arrived at Lily Pond around noon to find illegal campers had spent the night and morning there. They did not see me watch them take down their tent and pack up their belongings, and I was happy to leave matters at that. From time to time, they called and whistled for their absent dogs, which they had evidently allowed to roam freely. But when, after an hour, these pets failed to return, the couple drove off without them.

I was more than a little perturbed by this irresponsible behavior. What becomes of dogs and cats that are abandoned in state and national parks? And what impact do these domestic animals have on the wild creatures that have achieved some kind of homeostatic balance there? Surely, the owners would return to look for their animals, I told myself. I tried to remain calm, but I could not get the thought out of my mind that Huckleberry and Buttercup might soon become a meal. I could accept such a sad eventuality were it to be brought about by a wild predator, one native to the park, perhaps a coyote or a great horned owl. But I would not feel complacent about losing members of my colony to domestic dogs, animals forced to kill wildlife for food because of human dereliction.

While brooding over this depressing prospect, I spotted two Huskies bounding around the far end of the pond. They stopped at the place their owners had camped and sniffed the ground. I could see how uneasy they were at not finding anyone home, and I approached them with caution. My apprehension was unwarranted, however, for they were overjoyed to see a member of their adopted species, and after some coaxing on my part one even allowed me to slip my hand under his license-less collar. When I tried to lead him to my car, however, he growled, and I had to let him go.

What should I do about these abandoned dogs, I wondered. Since it was clear that I could not catch and transport them to an animal shelter by myself, I walked up to the road with the intention of flagging a park ranger or park policeman and asking for assistance. It was then I found Laurel.

She hadn't been dead long. Rigor mortis had not set in. A trail of blood on the pavement bore witness to the desperate effort she had made to reach roadside cover after having been hit. She was heading toward her new home after paying a visit to her birth pond when a motorized predator mowed her down.

How wasted a life. How terrible is the toll on animals caused by the automobile. The Urban Wildlife Research Center, using seven hundred independent studies, extrapolated that more than a million birds, mammals, and reptiles are killed *daily* on American roads. Even in this state park, where a forty-mile-an-hour limit has been instituted to prevent deer and driver fatalities, the carnage is terrible. Teenagers use straightaways to test their souped-up motors. And those who would respect the speed limit are often prodded to drive faster by impatient tailgaters.

I stroked the beaver's healthy fur. I mourned over the red two-year-old, and not just because I had a sense of who she was. No animal exists unto itself, and every creature that dies from causes extrinsic to the natural system, of which and for which it was born, weakens that system.* Moreover, the long investment by Lily and

*While it is true that scavengers feed on roadkills, in so doing, many become victims of the automobile themselves; thus the natural system, of which they are a significant piece, is still further diminished.

the Inspector General in bringing Laurel to a stage where she could at last create a suitable habitat in which to rear her own young—multiplied by a million dead animals a day—must also be added to the cost of road kills. How much natural selection and energy and experience and example and resource had gone into shaping Laurel? What provisions had been carried to her infant self, what biomass had been consumed, what risks had been taken, what care had been administered? And she had lived for two years and met every challenge nature had put to her. In so doing, she had proved herself superbly fit to pass the beaver's genetic code on to future generations. And now what had killed her was totally alien to this process of natural selection by which and for which evolution had shaped her. Death on the road is strictly a random event. Laurel's death had no meaning.

There are those who do not particularly value, have no sense of the uniqueness of every animal. There are wildlife experts whose single-minded aim is to maintain viable *populations* of species, who argue that the death of an individual creature is of no consequence, since most species are resilient and prolific and will reproduce at an increased rate in response to high mortality. Some species can do this, it is true, but not without cost to the systems of interacting flora and fauna that support them. Nothing is free, and inflated reproduction exacts a toll on habitat; for when turnover is speeded up, so is resource use. And when resources are too rapidly consumed, entropy occurs. Farmers and ranchers know this and rest and rotate their crops and pastures. Surely, nature did not grant species the ability to meet higher than ordinary mortality with elevated birth rates just so that man could justify his profligate consumption of wildlife. Compensatory breeding is a survival strategy that enables certain species to weather *cyclical* emergencies, not *constant* ones. A million unnatural deaths a day? Undoubtedly that figure is conservative. What of the hordes of uncounted dead—the mortally wounded animals who, like Laurel, crawl into roadside cover to die?

From whatever angle I viewed it, I perceived Laurel's death as a tragic waste, and my outrage made me want to blame somebody for it. I would like to have pointed my finger at the campers and their dogs. Had their presence somehow interfered with her movements? Why was she abroad on land in the middle of the day, anyway? Had she been caught too far from shore to make it back to Lily Pond through the danger zone they represented? But try as I would, I could not make that case to my own satisfaction. I wanted to blame the park administration. I was angry that speed-limit signs were so few and far between. Nowhere was information posted to explain to impatient drivers *why* they should drive with care, to remind them that animals use park roads to find mates, to go to water, to seek food. I was aware, however, that state parks operate on marginal budgets, and since in most states human rec-reation eats up the greater portion of what has been allocated, I understood why no such signs had been put up. I wanted to blame the driver who had struck down this beaver, but remembering that I, too, had once run over a raccoon, I could only sympathize with how that person may have felt. And yet responsibility for such unabated slaughter must reside with us, collectively and individ-ually, no matter how we avert our eyes and deny the gory evidence of our carelessness.

I did manage to stop a park patrolman and showed him the dead beaver and the vacated campsite and told him about the dogs, who by then had taken off, never to be seen again. That evening I sat at New Pond and watched the Skipper work all by himself on his dam, but my heart wasn't in it. He, of course, could not have known that his companion was dead. Yet the next night he aban-doned the place and moved back to Lily Pond. Animals are mys-terious.

Now I had only one site to keep tabs on, but even that task had become more complicated. As soon as Square Pond was completed, Lily and the Inspector General again expanded their waterworks, built two more beautiful adjoining pools above that one. These three contiguous ponds were set into a hill, one above the other, like terraces, so that by climbing over a dam, a beaver could exit one and enter another.

I was surprised at the amount of water contained by this chain of impoundments, especially in light of the fact that all were fed by such a small stream. What flow seeped under or over or through one dam was recaptured by the next, so that each pond lost no more water than it constantly received, and all three remained filled to the brim. At the same time, the feeder stream seeped on through all of them, ultimately flowing into and out of Lily Pond without being diminished. Thus the same volume of water descended to the bottom of the watershed as would have arrived there had no dams been built along its course. This point may be obvious to many readers, but I mention it because it would have escaped me had I not studied beaver dams. I am still amazed by the fact that beaver reservoirs do not deprive those who live downstream of water, but instead offer them more water in times of scarcity.*

Even while these three new ponds were inviting ducks and grebes and frogs and turtles to colonize them, much of New York state and New Jersey was drought-ridden. Every day dawned cloudless and hot with skies as blue as a morning glory. While beach-goers reveled in this run of fair weather, the huge reservoirs that supply one billion, eight-hundred million gallons of water to New York City every day, fell to an all-time low—less than 20 percent of their holding capacities. Photographs of their drained bottoms, hard and cracked like old china, were published daily in newspapers across the state, and night after night scenes of the worsening situation were shown on television. As a result, towns and cities throughout southern New York and northern New Jersey took steps to conserve what precious little potable water remained, and many types of consumption were banned. Suburban homeowners were forbidden to hose their lawns; city apartment dwellers were asked to fit devices to their shower heads to decrease flow; restaurateurs were ordered to stop serving unrequested water to customers. But these were Band-Aid treatments that did little to alter the fact that the area was running out of a vital resource. Was it a temporary condition or a signal of worse to come? A few news commentators observed that man's technology is not up to the task of inventing water, that man does not control nature, but is dependent upon her generosity.

Throughout this difficult time, the beavers were diligent about maintaining their dams. Only as much water as trickled into their pools did they permit to seep under or over or through their dams. Moreover, they added two inches to the one at Lily Pond without

*It's what happens when you place a catch basin under a leaky faucet. Eventually water spills over at the same rate that it leaks from the faucet, but now you have a basin full of water as well as a constant drip.

being prompted to do so by any noisy overflow. Most of their work on that dam took place underwater and involved the packing of mud against its base to prevent silent seepage. As a result, Lily Pond was maintained at its high-water mark, a mineral stain on the boulders that encircled it.

Now Lily Pond became a kind of wildlife spa, attracting more than its usual numbers of assorted creatures. The wet, lush marsh at its inlet not only sheltered resident mallards and the Canada geese who had nested there in the spring, but a tireless mink, who undulated about its ragged edges in search of muskrat houses to invade, a family of black ducks, a pair of little green herons, and a marsh hawk. And everywhere on the pond the bullfrog population thrived and celebrated its success loudly. And at dusk, the sky came alive with fluttering bats who preyed on the insect swarms that hatched off the water during the day. And along the shore tiny red efts hid under damp leaf litter and waited for the moment when they would be transformed into swimming newts and return to live in the pond that the beavers had made.

Sometimes I crossed the road to look at the abandoned water-works there. The New Pond dam, in want of upkeep, had rapidly deteriorated, causing the recently flooded valley to drain. Now no frogs concertized and few birds chased insects. How quickly the place had changed. I looked for sign of beaver, in the hope that some wanderer would come upon the place and put it back in order.

One day, to my surprise, I found a half-dozen scent mounds, conspicuously placed on tall hummocks. They were as large as any I had ever seen and fresh enough so that even my nose picked up their spicy aroma. Who had been here? Was it the Skipper? Had he returned in the dead of night to advertise for a mate?

The idea seemed preposterous. By now the Skipper seemed a permanent resident of Lily Pond. Whatever mood had come upon him in April, impelling him to seek a life of his own elsewhere, appeared to have subsided. All summer he resided in his parents' lodge, worked on their dams, dredged their channels, harvested their lilies, and disported with their youngest offspring, Huckle-berry and Buttercup. It looked to me as though he would remain at his birth pond forever.

When the leaves began to turn I stopped looking for beavers at

Aquatic life thrives at a brimful Lily Pond during a severe regional drought that depletes and devastates man-made reservoirs.

New Pond. It was then that the place, like an unwatched kettle, began to percolate with activity. In early October I made a perfunctory check of the site and found, to my astonishment, that the dam was being rebuilt and water was backing up into the little valley.

I could hardly wait for the gnome or gnomes responsible for all this renovation to appear, and I parked myself high on a steep slope to watch for them. They did not keep me in suspense for long. Soon a big brown beaver—was it the Skipper?—heaved himself over the dam and dived into what was fast becoming New Pond again. A moment later, a small brown beaver, who looked to be but a yearling, appeared and swam over to greet him. The two touched noses, then broke apart, and each went about the business of gathering sticks and adding them to the dam.

Who were these beavers?

The big one, I would soon conclude, had to be the Skipper, for from that night on, I never saw him at Lily Pond.

And the other? The Skipper's Second Mate, of course.

How had these two beavers found each other? Had a precocious

146

yearling (an émigrée before her time) just happened upon the spicy advertisements the Skipper had posted about the place? What took place when they met? How had they overcome *Castor canadensis'* innate tendency to avoid alien beavers, even be combative toward nonfamily? How had they progressed from casual contact to a tacit agreement to share property and lodgings with one another for life?

I ask these questions seriously, for beavers make these moves, form lifetime alliances, months before mating season casts its heady spell on both the sexes. Beaver pairing is based on an attraction that is as mysterious as it is compelling, one that is unrelated to any immediate urge to copulate.

I watched the two work until darkness obliterated their lumpy shapes. Afterward, as I drove home, I marveled at how full of unexpected turns and surprises beaver watching can be. Suddenly New Pond was coming to life again. Soon many animals would come to live there. And all because at some time in the recent past, while I wasn't looking, the Skipper had found a mate.

That night it rained.

Chapter Eighteen

That fall I cut down a birch tree and "planted" it in a hole I bored at Lily Pond, and the beavers eagerly felled and ate it. Why they fed on this tree while ignoring all those rooted about their pond was a mystery. One thing, however, was now clear; the beavers here had no need to lay in woody provisions for winter. Lily rhizomes would supply all their needs while they were sealed under ice. By what mysterious means they registered this fact eluded me. What could possibly inform them that enough feed to sustain them for three months lay buried in the bottom muck of their pond? For they seemed oblivious to signs of approaching freeze and scarcely bothered to cache food at all, only a few brushy items in an amount that could not hold six wintering beavers for a week. I would have expected that the impulse to cut and store winter food would be primal, encoded, inviolable, automatically activated by lengthening nights or cold weather or some such cue. It seemed to me that the process of natural selection ought long ago to have eliminated beavers who failed to make preparations for winter. Obviously, I was missing something here.

Shortly thereafter I inserted another cut tree in the flagpole-size hole I had made below my viewing rock. After packing dirt and stones around it to prevent its being pulled out by some smart beaver who discovered it wasn't rooted, I climbed up on my rock and waited.

Although I had seen Hope Buyukmihci use this trick to bring beavers up on her shore, I was surprised when it worked for me.

As soon as the beavers came out of their lodge that evening, they swam directly to the cove and eyed the "new growth" with what could only be described as keen interest. Do beavers take note of every change in the configuration of their shoreline? Or had they located the new tree by scent? Whichever, for half an hour they paddled back and forth in front of the miracle that had, like Jack's beanstalk, materialized while they slept, and each time they passed they fixed their gaze on its leafy crown. I hoped they were working up courage to perform an act I wanted to observe at close range, that is, the felling of a tree.

It was Lily who finally waddled up the bank, approached the birch, and pressed her nose flat against its trunk. For several seconds she remained in that position. Was she assessing the tree's health? Beavers, I later came to see, do not cut dead or diseased timber. They must, therefore, possess some means of determining a tree's health. Apparently, the particular birch I had brought passed muster, for Lily abruptly dug her bright orange incisors into its bark and pulled hard until a vertical shaving, three inches long, tore away. This scrap she promptly let fall from her mouth. Then began a rapidfire staccato of tooth against wood, a sound I would ever after recognize as that of a beaver felling a tree.

The process was quite different from what I had imagined. Lily did not simply gnaw on the tree willy-nilly; instead she tipped her head sideways, inserted her upper incisors crosswise into the tree trunk, then using these as anchors, began chopping with her lowers until they were well wedged in the wood. At that point, she tugged hard and stripped from the trunk what material her teeth gripped. After letting this scrap drop to the ground, she repeated the process. Thus, instead of chewing down the tree, as one might imagine, she chiseled and ripped it, bit by bit, and in so doing left a pile of sweet-smelling wood chips at its base. Only when one of these pieces got stuck in her mouth did she interrupt her nonstop chipping to fish it out with her black satiny fingers.

When the birch trunk, which measured slightly more than two inches in diameter, was nearly severed, it emitted a curious, creaking sound, and at this cue, Lily scrambled for cover—took off at a gallop, like a rotund pony bolting the starting gate. She made it to water just moments before the tree toppled, and by the time its

Above: Lily gives this tree a scent-check to determine if it is worth cutting. Below: To cut a tree, a beaver turns his head sideways and inserts his teeth at right angles to the trunk.

150

crown smacked the pond's surface, she was well in the clear. The entire felling operation had lasted but a few minutes, an astounding fact considering how much time I had spent cutting through the same trunk with a saw.

It soon became evident that the other members of the colony had not been oblivious to what Lily was up to. Before the expanding rings of water created by the tree's impact ran out of momentum, all five had joined her and were hard at work clipping and eating branches. Within an hour, the entire crown of the tree had been consumed.

While watching this communal feast I saw that the spring-born kits, Huckleberry and Buttercup, being considerably older for the season than had been the previous litter, were treated with less deference by their parents. Their wheedling seldom obtained them a branch from either parent. The Inspector General, in fact, grew quite intolerant of their incessant importuning and huffed and made mock lunges at the youngsters whenever they tried to snatch food he was enjoying. He was right, of course, for there was enough for all, and eventually, after they failed to obtain sticks from their elders, Huckleberry and Buttercup clipped their own from the crown of the felled tree.

By contrast, during the previous fall, the same plaintive sounds emitted by kits at an earlier stage of development had readily extracted food from parents and yearlings. Then Lily had even allowed the infants to share branchs she was feeding on. And though the Inspector General and Laurel and the Skipper had not been quite that solicitous, they had on occasion climbed up on shore to obtain branches for the beseeching infants.

Now Lily, in response to an impulse to hoard, towed a few birch boughs to the colony's skimpy cache. These, at least, would supply a bit of wood for the beavers to gnaw on during their winter confinement and so keep their teeth down. Beavers, being members of the family Rodentia, must gnaw on hard objects from time to time, for their incisors never stop growing from the root and require constant filing. Should a beaver be denied opportunity to chew, his front teeth become so long he can't close his mouth.

Meanwhile, across the road, the Skipper and Second Mate were now hard at work on an adjunct pond, downstream from their

After Lily fells the tree, every member of the colony picnics on it.

major dam. This new waterwork granted them access to many more shrubs and trees, food they would soon need for winter storage, for they did not enjoy the luxury of a crop of edible water-lily rhizomes buried in their pond bottom. Work on that project, however, held up construction of their living quarters, which was not begun until the third week in November. The site they chose for it, in the middle of New Pond, was unusual; it allowed them to incorporate a live bush into the lodge's west wall and to make use of a boulder to reinforce its east face. The end result was not only original, but good camouflage, for unless a person knew beforehand just what he was looking at, he might not make out the outlines of a beaver house, so obscured was it by stone and high-bush blueberry.

By now the region had soaked up several rainfalls, but the effects of the drought were still being felt, and the Skipper and Second Mate were assiduous in their efforts to capture more water. Each time the pond filled, they raised their dams' crests by a couple of inches. At this late date, however, it seemed unlikely that they would achieve sufficient depth to permit them far-ranging movement under ice cover. I tried not to worry about this. Perhaps they would have no need to travel any distance from their lodge during winter. Perhaps they would store enough food beside their entryways to make such forays about their pond unnecessary.

By late November, the New Pond lodge was essentially completed, although the newly paired beavers continued to plaster it with mud. They worked hard and long at this task, but with poor results. I concluded that the pond floor had not softened much since John and I had had a go at it with our hoes. Still, it was essential that they obtain their insulating material from a place beside the lodge, for just there they would need a deep pit in which to stash winter food. (Another example of the beaver performing two tasks at once.) Yet, despite steadfast efforts, they did not win the race with winter. The pond froze over before a single stick of food had been put away.

"It's only the first week in December," John reassured me by phone. "There will be a thaw, for sure. It's much too early in the season to take a freeze seriously."

Okay, I told myself. The Skipper and Second Mate are fat. They'll live for awhile on the food they've stored in their own bodies. As soon as there's a thaw, they'll get busy and lay in enough provisions to last them until spring.

But temperatures did not climb out of the teens, and by mid-December the ice cover was five inches thick. Now I could walk across it to inspect the new lodge close-up. Standing beside that rustic dwelling and listening to the beavers converse was like eavesdropping on a family of trolls. It was reassuring to know that the pair was still alive. But how long could they last without a stick of food stored? And what, if anything, could be done to help them now? I looked forward to the weekend, when John would come down from Massachusetts, but he offered me little solace. In fact, I had difficulty convincing him that the animals were actually in trouble.

"You say they're trapped in their lodge without food. Wild beavers wouldn't make such a mistake. Even if the water around their lodge has frozen to the bottom, I'm sure they got out while the getting was good and are living somewhere else now," he insisted.

The next morning we walked around New Pond and found a bank burrow that opened onto a beaver-dredged, still navigable channel. Bubbles, embedded in the channel's ice cover, bore witness to the fact that animals had recently been using it.

"There, you see, they're living in this burrow," John pointed out, "and these bubbles prove they are able to get around the pond. So long as they have freedom of movement, they'll find roots and things to eat on the pond bottom."

It was one of those rare occasions when we argued.

"That's a muskrat burrow. There aren't any beavers living there. And just because the beavers made that channel doesn't mean they are using it now. Muskrats made those bubbles. The beavers are frozen in their lodge, where the ice is so thick they aren't able to go anyplace."

"If that's true, and I don't believe it, just what do you expect to do about it?" John asked. "Break down their lodge and feed them? If we were to do that, they'd freeze to death. Try to remember how you worried about the Lily Pond beavers last winter, and in the end you found out they had food resources you didn't know anything about. I'm sure these beavers do, too."

I could not rebut this argument. Having been proved wrong about the Lily Pond colony, I had lost all credibility on the subject of wintering beavers. But just then nature joined my case: a muskrat shot out of the bank burrow and traveled swiftly under the channel ice upon which we were standing.

John mumbled something like, "Score one for you," and I seized the advantage and made a deal with him: We would stand by the beaver lodge and listen for proof that the Skipper and Second Mate were indeed inside. Even should this prove the case, however, I would stop fretting about them providing that their voices sounded okay to John.

In point of fact, the idea of keeping two beavers alive all winter did not appeal to me any more than it did to John. I doubted it could even be done. What cuttings I had contributed to the Lily Pond beavers had been carried by them to their food cache. Here at New Pond, however, the situation was different. The Skipper and his Second Mate were in real trouble, and I could argue both sides of the question of whether or not to give them aid. I was, in fact, quite proficient at making a case for letting nature take her course. Ought not animals who fail to stock their winter larders be eliminated from the race for survival? And what about natural population regulation? Although beavers reproduce relatively slowly, many more young are born than can eventually find niches in which to live. A percentage must die.

But Laurel and the Skipper *had* found a niche in which to live. And were it not for a highway accident that held up work on the pond—an incident that was alien to the natural selection process —those two might have met their timetable and laid in winter feed. Moreover, they would have stayed ahead of the drought and captured sufficient water to assure that their pond did not freeze to the bottom. Nature had not marked those animals as losers in the competition for survival; man's activities had undermined the selection process. And since *Homo sapiens* had had such a negative impact on the inhabitants of this pond, what would be wrong with man making amends?

John and I studied the thick ice around the lodge.

"How would we get food to them anyway? That is, if they are still alive?" I asked.

"I could chop a hole in the ice with an ax and then we could stuff sticks into it, I expect," John replied.

We inspected the vent hole at the top of the lodge and found that the snow cover there had crystallized, a good sign that living, breathing animals were inside. Then, after a wait of perhaps forty-five minutes, we heard a faint whine. The voice sounded weak, and I could see that John was now becoming interested in doing something to help the animals.

"This is the deepest spot, right here," he announced, after walking around the lodge a few times and peering through the ice.

"I can cut some black birch from my yard," I prompted.

Now any lingering reservation about what we were about to do evaporated. Surely a single act of compassion on our part would not throw the park population out of kilter or even begin to outweigh man's careless and sometimes deliberate destruction of the beaver.

That night we waited beside the large bundle of branches we had inserted in a hole John chopped in the ice. The pond, we discovered, was so shallow that the upper portion of the boughs protruded above its frozen surface. Just to find a spot where water still flowed beneath the thick ice, thus allowing the beavers access to our offering, John had had to break through in several places. As we stood in the dying light, we felt the temperature drop suddenly and had to stomp our feet to keep warm. The cold spell was not about to abate. Then, just as we were ready to give up for the night, we heard a glug, and slowly one of the branches began to tremble and wriggle and slowly disappear into the water.

"They've found the food," I whispered.

More glugs told us an animal had entered the lodge with the manna he had discovered. Then two beaver voices intoned together—"uh-UH-uh-UH-uh-UH-uh," followed by the rhythmic scraping sounds of beaver teeth chiseling bark.

Overhead, the moon wore a halo, a sign that snow would fall during the night. Already the hole containing our offering was refreezing. Soon only the lower half of the sticks we had inserted in it would be available to the sealed-in beavers.

"Even so, that ought to hold them for a week," John said. "And we'll put in a lot more after Christmas. Then in January, we're sure to see a thaw. After that they're on their own."

We listened awhile longer to the beavers' contented murmurings. Then another glug told us our offering was being visited again.

I was delighted. The animals had food, and knowing this was going to make my own dinner taste a lot better. We headed for my cabin, and as we drove through the park I mused aloud: "What on earth do you suppose those two animals make of all this?" It was a rhetorical question, but John picked up on it.

"I couldn't say," he replied with a laugh, "but given the bind they were in, I suspect their conversation tonight is philosophical. In fact, at one point I believe I heard one say to the other: 'Yes, now at last we have convincing proof that there is a god'."

Chapter Nineteen

Between Christmas and New Year's, John chopped another hole in the frozen pond and we stuffed more branches into it. This time we lifted out all the ice John had bashed to bits, rather than shove it under the ice cap, for the crawl space on the pond bottom was so tight I feared ice debris would block the beavers' passage to the food we had brought.

Almost as soon as we inserted a branch into this opening a muskrat appeared and stole it. He was extraordinarily quick. Nevertheless, we spied his shadowy form moving under the ice, and heard him carry his booty to an apartment he had made in the wall of the beaver lodge.

So we were feeding muskrats as well as beavers! This was not to be the only time I would discover muskrat lodgers in beaver quarters. The following spring I observed that distant relative of *Castor canadensis* living in the walls of the Lily Pond lodge, just at a time when new kits were beginning to make their evening appearances. Even more extraordinary was a discovery the following winter that beavers were making use of a muskrat house! We speculated that the little reed hut was serving as a pit stop for the beavers, a place where they could refill their lungs with air. Now seeing a muskrat disappear with a part of our handout made us glad we had cut such a hefty bundle of aspen for the New Pond beavers.

That week I paid two visits to Lily Pond to listen for news of the six gnomes who were wintering inside the snug lodge there.

They sounded in fine fettle. I heard them come and go and converse with one another in contented tones, as they had done during the previous winter. At one point, a beaver swam under the ice to the end of the pond and gnawed a hole in the dam, thus allowing water to run out. The existence of this spillway lowered the pond by a couple of inches, creating a space between the ice cap and the water surface that provided the beavers a handy source of air.

After New Year's Day, John returned to Massachusetts and I was on my own. Every day I looked for sign that the cold spell would end. Every evening I visited New Pond to listen for news that the beavers were surviving.

The hole into which we had inserted our last food drop had resealed itself, but the ice that formed over it was of a different quality, less dense and only half as thick as elsewhere on the pond. Whether this was because it had once been broken or was a result of beaver activity at the site, I cannot say. In any case, I found that even I, without too much exertion and with the help of a crowbar, could punch it out. Then by probing underwater with a long stick, I was able to determine how much of the feed that we had cached still remained. When all was gone, I brought produce from my refrigerator—apples, carrots, sweet potatoes, yams, and lettuce. The apples were worse than useless. Try as I would, I could not sink them. I poked at bobbing Macintoshes for half an hour, but failed to push a single one under the thick ice cover. In less than an hour all had become embedded in newly formed ice. Carrots and sweet potatoes, on the other hand, were easy to plunge into the crawl space at the pond bottom, and the beavers evidently had no trouble finding them. After a short wait, I heard the crunchy sound of munching, quite a different tune than the scratch, scratch, scratch of tooth against wood.

January passed without sign of a thaw. Several times I pounded out the same hole and worked sweet potatoes and yams under the ice. Before leaving, I always placed cardboard over the opening to insulate it from cold air and make it easier to punch out again. Nevertheless, by February the beavers were in serious trouble. Their voices sounded faint, and occasional moans wafted through the lodge chimney. So solid and thick was the pond's frozen cover that a person could have driven a truck on it. I was no longer able to punch open the feeding hole. Not that there was much point in my doing so, for the underlying crawl space had shrunk to four inches or less. Beavers would not be able to maneuver about under such a low ceiling.

I recalled reading of a similar situation in a book titled *In Beaver World*, by Enos Mills, a naturalist who watched beavers in Colorado at the turn of the century. During a particularly hard winter Mills and two friends became concerned over the fate of a lone beaver when they noticed that the underwater pit into which the animal had sunk his winter feed had frozen to the bottom. When it became apparent to them that all entryways to the lodge were impassable, they decided to do something about the situation. Mills described their efforts as follows:

> We broke through the frozen walls of the house and crawled in. The old fellow was still alive, though greatly emaciated. For some time—I know not how long—he had subsisted on the wood and the bark of some green sticks, which had been built into an addition of the house during the autumn. We cut several green aspens into short lengths and threw them into the house. The broken hole was then closed. The old fellow accepted these cheerfully. For six weeks aspens were occasionally thrown to him, and at the end of this time the spring warmth melted the deep snow . . . and the old fellow emerged into the water.

I refused to contemplate breaking into the New Pond lodge to deliver food to the Skipper and Second Mate. Beaver walls can be a foot thick and are weatherproofed with layers of hard mud to protect the animals from the elements. Our bungled attempt at repairing the dam we broke convinced me now that I would not be up to the job of patching a lodge. Instead, I appealed to Dan's seventy-eight-year-old father, Henry Pierson, and to a local friend, Tony Spadavecchia, for ideas. After hearing about the beavers' plight, Tony volunteered to cut through the ice with his chain saw, and Henry suggested we break up twigs into tiny bits, bring them to the pond in plastic garbage bags and pour this detritus into the hole Tony would saw open.

While Henry and I spent an entire day reducing aspen trees to two-inch lengths, we had to laugh at ourselves, for it seemed more than a little ridiculous to be cutting up branches and twigs for so notable a woodchopper as the beaver. Moreover, we had to wonder if what we were doing was really necessary.

Late that afternoon, however, all our doubts vanished when Tony's power saw struck bottom. A geyser of black muck spattered high into the air.

"There's no water at all under this ice," Tony yelled above the whine of his saw.

After cutting to the bottom in several places, however, he finally located a spot beneath which a tiny crawl space still existed. After removing the block of ice he carved out with his blade, Henry and I filled the resulting hole with aspen detritus and packed it down by jumping on it. Eventually, it was as stuffed as a mattress.

After such a day we were too exhausted to remain at the pond to listen for evidence that our efforts had brought relief to the beavers. The following night, however, I heard the glad tune of gnawing coming from inside the lodge and so was able to report to Henry and Tony that our relief drop had reached its mark.

Over the next few weeks I punched open Tony's hole and refilled it with feed again and again. Mostly I dumped in produce—five-pound bags of sweet potatoes or yams. Once I tossed in turnips. Who knows what beavers eat? Judging by the Lily Pond colony's weakness for lily rhizomes, I figured any kind of root or tuber would do.

New Pond did not thaw completely until the end of March, three weeks after Lily Pond was ice-free. During most of that month, while Lily and the Inspector General were porpoising about on open water, the Skipper and Second Mate still required my help. By then I understood why. New Pond was situated between two steep hills and was shaded during most of the day. As a result, the place was so cold that John proposed we rename it Refrigerator Pond. When water did at last begin to seep up along the edges of its ice cover, I stopped the food deliveries. In all I had made fourteen.

My first sighting of the Skipper and Second Mate occurred in early April. I had not expected to find them in such good shape. Their coats were red-tinted (from so much carotene?), shiny, and dense. A diet of sweet potatoes had apparently not hurt them.

Immediately they set to work repairing their auxiliary pond. When satisfied with it, they dredged two long canals farther downstream, bisecting a meadow that had in times past been a large beaver pond. I hoped the pair would convert the place into a waterwork again, for it could serve them well. A plethora of vegetation grew around its edges, and no surrounding hills cast long shadows over it. Come winter, it would not turn into a block of ice. I named this potential pond-site Valhalla, and followed the stream that exited it for more than a mile. Along its course, I discovered five more abandoned beaver ponds that were ripe for recolonization. Should the Skipper and Second Mate produce a number of offspring, there was ample room for expansion on this watershed.

One evening, shortly after ice-out, the pair put on a show for me. In fact they behaved so coquettishly it struck me they were engaged in some kind of sexual foreplay. For want of adequate underice space in which to copulate during February and March (when beavers normally mate), the female had not become impregnated and now had come into estrous again. I watched the pair tumble over and under one another while both grasped the same stick with their teeth. What ultimately took place I did not see, for it happened underwater, but in early August the outcome was visible: two fat-faced kits. I named one Sweet Potato and the other Yam.*

I had high expectations for the success of this new colony. There was plenty of willow in New Pond to support the little family over summer, and it looked as though the pair was shoring up enough water at Valhalla to enable them to winter there. In this effort, they were not handicapped by a delayed start, as they had been the previous year. Nor did they have to battle drought. But would they know enough to build living quarters there? Or were they attached to the beautiful lodge they had created at New Pond? Would they remember the Arctic conditions they had endured in that place? Or had the experience of receiving handouts made welfare beavers out of them? One cannot predict the ultimate outcome

*It was after the birth of these kits that I discovered the Skipper's nipples were swollen and that "he" was a "she." Second Mate then had to be the male of the pair. Shortly thereafter I received an autopsy report stating that Laurel (the Skipper's late sibling with whom she had homesteaded) had been a he, not a she. Always, if I wait long enough, truth will out.

of one's best intentions. Good deeds sometimes produce unfortunate ends. Perhaps I was now a slave to this beaver colony, and they to me.

I need not have worried on that score, however. The little colony worked hard and expanded their waterworks, grew fat on willow, and then abandoned the entire watershed. I do not know the exact day they departed, for I spent most of September observing the colony at Lily Pond. But one evening around the autumnal equinox, John and I visited New Pond and sensed something strange about the place, a disturbing stillness.

"Beavers don't live here anymore," John finally said.

"They must be sleeping late," I replied. "They'll be out soon."

But New Pond seemed as haunted as the deserted cabin of a long-gone prospector. Where could they have gone? And all together?

After searching the lower part of the watershed for sign of them, we returned and sat on the steep bank beside the pond. There we waited far into the night, straining to catch sight of a beaver wake, straining to hear the scratch, scratch, scratch of beaver teeth on wood. By midnight, we knew with certainty that the colony had departed.

The following day I pored over the literature to learn if entire colonies pick up and move to new locations en masse. Harry Hodgdon recorded the arrival of fifteen such intact family groups to his Quabbin study area over a period of six years. Moreover, Hodgdon observed that eleven family groups had departed the vicinity during the same period. These arrivals and departures occurred either in early spring before kits were born, or in late summer after the offspring were two months old. It would seem then that beaver parents wait until their babies are strong enough to withstand the rigors of travel before embarking on a major move.

For several days I looked up and down the watershed for the New Pond colony, but found no evidence of them. Where might they have gone? I did not suspect foul play. Beaver fur is not prime so early in fall, and trappers know it.

It wasn't until December that I discovered their whereabouts, and then quite inadvertently while exploring the linked pools that the Lily Pond beavers had created above Square Pond. The top one of these waterworks received inflow through a culvert that ran under a busy park thoroughfare. As the level of this uppermost pool rose, water backed through that culvert and flooded a natural basin on the other side of the road. There, hidden by alder and dense stands of phragmites, was a handsome beaver lodge.

It took some doing to determine just who inhabited this place, so well concealed were its residents by swamp vegetaton. In time I did count two adults of slightly different sizes and two fat-faced kits—a perfect match for my missing family. Simultaneously, John remained at Lily Pond and accounted for all the beavers there, thus eliminating the possibility that I had included any of them in my tally. Still, I was not fully convinced I had found the New Pond colony. Not until I discovered a beaver-worn overland trail connecting the works with New Pond did I believe my good luck. I named the place Top Pond.

Needless to say, I was overjoyed to make contact with these four animals again. Further observations of them would help answer questions raised by their move. What, for example, had prompted them to abandon New Pond where plenty of feed still remained? Were they motivated by a recollection of the predicament they had found themselves in during the previous winter? But if their memories were that good, would they not recall that imprisonment had not prevented food from being delivered to their door? Would they not expect that miracle to be repeated? Surely memory and experience help determine how and even if a beaver colony prepares for winter. Otherwise, the Lily Pond beavers would cut and lay in large food caches every fall, despite the fact that plenty of lily rhizomes lay buried in their pond bottom. Instead, they felled almost no trees and cached only a small amount of underbrush. Now I wanted to see if the Skipper and Second Mate, lacking a buried treasure of lily roots and having once experienced life on the dole, would lay in winter food.

To my great relief, they did. Without help from their offspring, the pair managed to cut and cache scores of alder, several large oaks, and a few swamp maples—more than enough feed to sustain the family over winter. Thus they demonstrated that being on the skids need not be habit-forming, even for beavers. By helping them, I had not simply forestalled the inevitable demise of a pair of losers: this pair functioned superbly.

Chapter Twenty

W hile I was laboring to deliver food to two incarcerated beavers in the New Pond lodge, the colony at Lily Pond wintered well, gorging on lily rhizomes in such excess that by spring the shoreline was littered with partially eaten roots, each one bearing witness to the beavers' profligacy. Many appeared to have been hardly nibbled before being cast away. Such behavior on the part of wild animals is not unusual and of small importance, though human critics often think otherwise. In actual fact, little if anything in a natural system is lost. What food is not consumed by the animal that harvests it will be discovered by another or will be returned to the earth in the form of nutrients. On the other hand, human activities do indeed contribute to the permanent degradation of natural systems; for example, the loss of valuable topsoil by strip mining, the pollution of water tables by chemical dumpers, the destruction of entire forests by acid rain. By contrast, even when animals overharvest the very resource that is vital to their survival, the effect simply hastens their departure from the site, thus insuring that plant food a period of recovery. Realizing this, I eyed the plethora of partly eaten roots that had washed ashore not judgmentally, but with regret. I did not want to lose my Lily Pond colony. At the same time, I was happy that all six beavers had fared so well over winter, for they emerged from the lodge fatter than ever and in robust health.

Huckleberry and Buttercup, who had entered the winter as babies, appeared so grown in spring that I had difficulty distinguishing them from their older siblings, Blossom and Lotus. From any distance, all four looked the same size. Moreover, the lighter beaver in each age class had become darker, just to confuse me further. I saw no recourse but to lure the entire family to the cove and inspect them all anew. So once again I placed aspen branches in the water and retired to my viewing rock to await the beavers' evening appearance.

At close range I was able to detect some previously unnoticed characteristics. Blossom, for example, had at some time over the winter incurred a minor injury, a split on the left side of his tail tip. Since beavers frequently hold their tails aloft while feeding, this gave me a long-distance marker by which to distinguish him from my other dark beaver, Huckleberry. I expected that Lotus would soon turn blond, as she had done the previous spring; meanwhile, I discovered that she and Buttercup carried an indelible marker for their predisposition to fade, namely, pink claws. By contrast, the claws of the other beavers were either dark gray or yellowish gray. Identifying the Inspector General was no trick. No beaver came close to him in size. And Lily, wondrous creature, retained that quality of expression that was uniquely hers and that never failed to incite in me the urge to reach out and touch her. Her muzzle had become more grizzled over the winter, and now I wondered how old she might be.

Few wild beavers live even as long as a decade, but in 1966 one female whose teeth and eye lenses suggested to researchers that she had attained twenty and one half years was trapped in Maryland. Though captive beavers have been known to surpass even that great age, this does seem to be the record for beavers living wild; the Maryland study found no other animal that had survived beyond twelve. Lily, I suspected, had already lived well past the norm.

Once I had scrutinized the Lily Pond colony at close range, I was again able to pick up individual characteristics at a distance through binoculars. Though all age classes interacted with each other, littermates, I noticed, preferred one another's company to that of older or younger family members. After being apart for a time, same-age siblings would approach and greet with loud vocalization and by touching noses. Sometimes they would swim apart, one traveling clockwise, the other counterclockwise, each

During the four years I observed the colony, Lily showed signs of aging.

tracing a perfect semi-circle until they met again. Sometimes they would engage in a round of play or dive alongside one another while traveling about the pond.

On rare occasions these water games got out of hand and led to shoving matches. Like Japanese wrestlers, the contenders would square off, grip one another's loose ruff with their black satiny hands, and then drive forward with all their might until the stronger one propelled the weaker backward into deep water. Breast-to-breast, cheek-to-cheek, heads tilted skyward, eyes rolled upward so that only membranes showed, their resemblance to Samurai warriors was uncanny, both in bodily shape and in the martial strategies they employed. They inflicted no wounds; theirs was a contest of strength, not an outlet for vengeance. Nor was the animal who was momentarily being bested automatically defeated. As the prevailing beaver tired, the receding one summoned strength to halt his ignominious retreat. Then he or she would drive forward, forcing the other to back paddle. Thus locked together, the grunting beavers advanced and retreated, advanced and retreated, advanced and retreated until at last one gave up. Giving up was all that was necessary to satisfy the victor. The vanquished animal was not required to grovel or make a show of his submission, as does a defeated wolf. No blood was shed, no grudges festered as an outcome of these beaver contests. Afterward the combatants swam off together and behaved as if nothing untoward had occurred.

I watched these matches many times and thought a lot about them. Any species that possesses sharp teeth with which to obtain food must avoid using those hazardous tools against its kind, for doing so could bring about its extinction. Moreover, it is certainly not in the interest of an individual animal to kill a close relative whose hereditary make-up (being similar to his own) offers a backup means by which his own genetic material may be propagated. Finally, only a colony that is able to live in peace is assured the help of many hands and jaws in the creation and maintenance of its vital waterworks. Thus it is not surprising that the beaver has evolved strong inhibitions against biting, together with a ritualistic means by which to safely settle disputes.

That is not to say, however, that push matches never escalate into out-and-out warfare. Though biting is under strong inhibition, this restraint can be overruled—especially when the animals who come into conflict are not of the same family. Although strangers are, as a rule, ignored by resident animals and allowed to hurry through already claimed territory, should any transient linger too long, he may find his hindquarters under attack. And if the trespasser then stands his ground, a wrestling match will erupt that can rapidly intensify into serious combat. Lars Wilsson describes the process as follows:

> When on land, both animals rise on their hind legs using the tail for support and balance, grasp each others' skin with both hands and try to push each other backwards. . . . The nearer the animals are to the lodge of the territory, the fiercer is the fight, and finally the incisors are used as weapons. The bites are mainly directed against the opponent's hindquarters, and the trespasser may be killed by having its spine bitten off just above the root of the tail.

I never saw any of my beavers engage in that kind of fight. All their push matches ended peaceably, and many appeared to be mere expressions of playfulness. Like dogs, who employ the same stereotypical moves in a mock fight as when they truly join in battle, beavers appear to indulge in ritualized wrestling just for sport. Dorothy Richards interpreted every match she witnessed as nothing other than play. She recorded the behavior in two five-day-old kits whom she was rearing, between a mated pair, and by some half-grown residents of her pond. In *Beaversprite* she describes how a group of youngsters once paired off and put on a wrestling performance that reminded her of dancers doing the old-fashioned Bunny Hug.

I often failed to see what triggered the eruption of a wrestling match, but once I clearly did. The whole thing started when two beavers attempted to snitch a branch from a third one, who was blithely feeding on it in the water. They approached stealthily, like two submarines, and attacked from two directions at once. In a microsecond, one had seized the prize and was off with it by underwater route. The startled victim, upon realizing his branch was gone, turned on the luckless raider who had failed to obtain any part of the booty, and after a short tussle in the water, the two engaged in serious wrestling that lasted a full two minutes. Afterward peace was restored.

That spring Huckleberry and Buttercup seemed to grow before my eyes, and at any distance appeared to be big beavers, even though they had barely achieved yearling status. I attributed their large size to the colony's out-of-the-ordinary winter fare; bite-for-bite, the caloric content of lily rhizomes likely exceeds that of cambium-lined bark.

Now the two refused to give up dining on that food. Long after the crispy, stalk-like roots had given rise to equally edible pads and flowers, Huckleberry and Buttercup continued to dredge and eat them. In so doing, they not only depleted the beavers' winter staple, they eroded the colony's summer mainstay as well, for each rhizome they unearthed supported numerous long-stemmed leaves and flowers, which the youngsters let go to waste. For that matter, they rarely consumed an entire root. After taking a few bites from one they cast it aside and mined the muck for another. As I saw it,

Beavers settle disputes by holding a push match, a nonviolent test of strength similar to arm wrestling.

their addiction to this food threatened the colony's most renewable resource. Already I had observed a significant change in the look of the pond. Two years earlier, its entire surface had been blanketed by lilies; now those plants lay in matted patches here and there, scattered islands of jade and white surrounded by expanses of open water. And while this altered condition allowed me to spot beavers with greater ease than ever before, I had to wonder what the colony would use for food, come winter.

Of course, I had expected something like this to happen. Beavers do "eat themselves out of lodge and pond" and then must move elsewhere. Such is the fate of a creature whose way of life is profoundly bound up with biotic succession. Even so, I harbored a private hope that the colony would not prove equal to the task of uprooting a crop of water lilies that had reached such proportions. If indeed it did, such ability deserved recognition and ought to be put to use by those who clear lakes and ponds of the hard-to-control water lily. I remembered reading that a South American relative of the beaver, the nutria, was introduced into waterways in the South during the 1930s in the hope that it would perform the very task. Now I suspected a home-grown animal would have been a better candidate for the job.

By June the fat yearlings, Huckleberry and Buttercup, were performing work appropriate to youngsters of their age. Huckleberry became obsessive about adding material to the dam; no matter that the structure was in absolutely no need of repair. So absorbed did he become in diving for muck and transporting sticks to the structure's overbuilt crest that he seemed oblivious to all else, including my presence at three feet.

As for Lotus and Blossom, their interest in one another was hard for me to interpret. Were they engaged in a dominance contest of some kind? When not feeding—which is what beavers mostly do in spring—they converged on the pond to greet, nudge, dive, talk, and wrestle. I suspected that they, like the Skipper and Laurel, had become a bonded pair and might any day slip off together to found their own colony.

Meanwhile, the Inspector General was actively harvesting grass and carrying it into the lodge—a pretty good sign that newborn kits were inside; Lily was conspicuous by her frequent absences from the pond—a pretty good sign that she was spending time with infants; and try as I would, I could count but five beavers out of the lodge at any one time—a pretty good sign that one or another of the six was remaining behind to baby-sit.

Then disaster struck. It was June 20, the eve of the summer solstice. The weekend was to have been a special one. In all the year, no twilight lasts so long, no opportunity for beaver-viewing is so extended. John and I arranged to be at the pond by mid-afternoon on Friday so we could examine the upper waterworks before the animals woke up. Afterward we planned to spend most of that night observing the beavers.

After parking, we followed a short trail that led to the northwest corner of the dam; but even before we came to an elevated section along that footpath, at which point the pond becomes visible, we both knew something was terribly amiss. A menacing sound, the steady drone of fast-flowing water, galvanized us, and we broke into a run.

"It's the dam," I shouted.

As the pond loomed into view, we both stopped, stunned by

what we saw. Water was spewing forth through a break in the five-foot-high dam with stupendous force, carving an ever deepening, ever widening crater in the structure. The pond was sinking before our eyes.

I raced downhill toward the broken dam, falling twice in panic, and then edged across a still intact part of it to the very brink of the catapulting water. There I stood, helpless in the presence of such awesome power. Like an oil gusher, like a broken water-main, like a torrential flash flood, water surged past me with astonishing force. Below the dam, the land naturally fell away, allowing the released water to gain ever more momentum as it searched wildly for the shortest possible route to the sea. A fish, caught in the spate, swept by. And wood, large pieces of the dam, wood that the beavers had cut, years of their cuttings, washed past me and cascaded downhill.

I stood without moving, stupefied, rendered impotent by what was happening. Then I became aware of John moving toward the breach, struggling to carry a small boulder. Oh, he will staunch this flow, I thought. My spirits momentarily brightened. But the heavy stone he tossed catapulted past me, riding the torrent as if it were a beach ball.

We stared after it, tracked its swift course downhill with our eyes. What energy had been unleashed here. John spoke, and his words were barely audible above the roar.

"This pond will be gone in two hours."

He hadn't meant for me to hear.

"Oh, why? Why? Why has this happened?"

I expected no response. My question was a lament, a protest against the irreversibility of such an event—the cry of Job railing against an act of God. Yet my words alerted John to look for an answer, to take note of the ground. A man's track was evident everywhere around the break.

"Why, this is the work of a vandal! See here? The dam has been pried apart with a tool of some kind."

There it was. The evidence that we had almost neglected to look for.

Now, having connected the disaster to human perniciousness, rage acted as a catalyst in us, unleashing our power to act. It was

The dam has been vandalized. Rushing water erodes an ever-deepening channel, causing water to empty from the pond at an increasingly rapid rate.

quickly decided that I would drive seven miles to the nearest phone to notify park police, while John would remain behind to await their arrival, for should a radio car be dispatched from a nearby point, it likely would reach the pond before my return. Meanwhile, John would try to implement an idea he had.

"The pond is sinking fast," he explained. "Obviously, I can't place rocks directly in the breach, the water has too much force just there. What I will try to do is tie the severed dam together with an underwater stone wall that bows out into the pond. With luck the gathering force of the water, as it nears the drop off, will tighten the stones, rather than wash them away."

Then seeing hope in my face, he hastened to add:

"You understand that such an underwater wall will be so full of chinks it could not possibly stop the pond from draining. Still, if it slows down the process, we might buy some time. Then maybe the park rangers can figure out a way to drive in here and dump a load of rocks in the breach—or some such thing. Anyway, I can't think of anything else to try."

I didn't question the soundness of John's plan. Taking action, however futile, seemed better than standing by. I struggled to pick up a massive rock to assist in his ambitious scheme, but he chased me away.

"Go get the police," he ordered.

Halfway up the path, however, I thought of something I had to do first and turned back. Quickly, I photographed the craterlike break, shot a few pictures of John heaving stones into the pond, and documented on film the incriminating tracks we had discovered on the dam. I wanted evidence to convict whoever was responsible for this heinous crime. On behalf of the legion of wild creatures who would soon die or be set adrift because of a deliberate and vicious act of man, I would seek vindication.

Chapter Twenty-one

By the time I got back, two police officers were already on scene, standing beside John on what remained of the dam and shaking their heads over what they were seeing and hearing. As I approached, I overheard them conclude that nothing would stop the inexorable flow of water from the pond, that the huge break in the dam had been deliberately created by a vandal, and that the pond would be history by morning. So much for man's technology, I thought. We have the means at our command to make the planet uninhabitable with the press of a button, but we aren't able to save a beaver pond.

After introducing myself, I answered routine questions about the incident, then added a few facts about the value of beaver ponds and described the colony's current status.

"There are kits in the lodge," I said, "and they must remain there, because they are too young to maneuver about in water. But babies or no, once the beaver house is high and dry it will be abandoned. The colony must be assured a covering of water over its entryways as protection against predators. Where the family will go and what will become of the kits is anybody's guess."

Both men seemed moved by the beavers' predicament and expressed a desire to see whoever was responsible brought to justice. But they quickly added that they did not hold out much hope of catching the guilty party.

"Even if we nabbed the guy, it wouldn't be easy to get a conviction. We would need an eyewitness, and you people didn't actually see who did this. While the footprints and toolmarks leave little doubt in our minds that somebody deliberately broke the dam for some sick reason or other, we'd have trouble making a case in court with that kind of circumstantial evidence."

The deed then would go unavenged. Worse, the vandal would remain at large and might at any time drain other beaver ponds in the park.

"Be careful when you work here at night," one of the officers warned me. "This doesn't look like teenage high jinks. The fact that a single individual was involved suggests it was done out of malice. Someone may have a grudge against the park. Or possibly it was the work of a deranged mind."

The two men then assured me they would be on the lookout for anyone acting strangely, but added that they were understaffed to deal with all the problems that arise daily in such a large park. Even while they were speaking, a call came in over their walkie-talkies, and they had to rush off. A woman was being molested at a park campsite.

During all this time, John had made no mention of the underwater wall he had begun building before I left, and now I was eager to hear what had come of it. No rocks were visible above the surface of the pond, but it seemed to me that the torrent of water pouring through the break had lost some power.

"It's not moving as fast as it was," he conceded, "but the pond is still draining at a pretty impressive rate, and there's no way anybody's going to stop it. The pile of rocks I threw together out there is like a colander. It only serves as a breakwater to slow down the current, that's all. Let's face it, you can't hold back Niagara Falls with a fish net."

I had to agree. I was as convinced as John that the beavers would only try to stop the noisy spill of water flowing over the top of his jerry-built breakwater. Everything we had ever seen, heard, or read about beavers informed us that they would automatically act out this encoded response, however futile. Meanwhile, silent underwater leaks, such as those that were now inexorably draining the pond, would remain untended.

We climbed onto a big rock and studied the shoreline. High-water marks on bank boulders indicated that the water level had

John attempts to build an underwater rock-wall to check the powerful current and slow the escape of water.

176

dropped by more than two feet since our arrival. Soon the beaver lodge would stand on dry ground, like a tepee, with all doors wide open to passing predators. Already the tops of its entryways peeped above the water line, like three dark moons rising. And given the tremendous force at the point of spillage, we came to the conclusion that the beavers would not be able to repair the dam until the pond was completely drained. Should they try, the current would carry them over the falls.

During the next half hour the pond continued to sink, gradually revealing the uppermost stones in John's wall. I was now able to understand how his makeshift jetty functioned. From the downstream side, I could see water gushing through chinks and crevices in it, to say nothing of the torrent that poured over its rim. Still the bowed wall reconnected the severed dam and did indeed reduce the force and size of the outflow. John seemed surprised that his unmortared masonry was holding up at all. Privately, he had expected it to crumble under so much pressure.

Suddenly my heart began to pound. A beaver had emerged from the lodge and was swimming toward the dam at top speed. By the time I pointed, the animal had dipped underwater and did not surface again until a few feet in front of us. It was the Inspector General and he appeared agitated. Ignoring our presence, he surveyed the broken dam with a wild eye, swam back and forth alongside the porous wall and took in the dreadful scene. Then he sped off to the far shore, where he felled a large laurel shrub with amazing dispatch.

"He's going to try to do the impossible," John said.

"Yes, and he's going to wear himself out failing. It's too bad there are no fallen branches around here. I wish we'd thought to gather some from across the road before he woke up." (Illegal campers had picked the banks of Lily Pond clean of firewood.)

As we watched, the big beaver hauled a six-foot shrub into the water and began transporting it to the break. The scene reminded me of the witches' prophecy in *Macbeth*. "Here comes Birnam Wood," I said, for the bush was in full bloom and traveling without visible sign of a beaver beneath its dense foliage. At John's submerged wall, the apparition idled for a moment, then its voluminous crown

The Inspector General discovers the dam break and makes a fruitless attempt to stop the outpouring with a laurel cutting.

178

rose into the air and tipped over the rim of that rockpile. For a moment, I thought the huge cutting would be carried downstream by the torrent of water cascading over, under, and through it. But the beaver held on to its short stem with his teeth and, using his front handlike paws, wedged its butt end in between two top stones, thus hanging it over the backside of the underwater wall.

"That's not going to do much good in this kind of flow," John commented.

The beaver must have come to the same conclusion, for he made only one more trip to the far shore to cut, tow, and anchor a second laurel branch before abandoning that tactic in favor of another. Now, to our amazement, he turned his attention to underwater leakage. To plug the myriad crannies in the rock pile through which enormous volumes of water were escaping, he uprooted whole lily plants, higgledy-piggledy to the right, left, back, and front of him and he used these as caulking material.

"He's going right to the heart of the problem!" John said. "I wouldn't have believed this if I hadn't seen it. According to theory, he should pile all that stuff on the crest of the dam to put out the noise, so to speak. Instead, he's using it underwater where it's most needed, and it's a sure bet he can't hear water escape down there. So what goes?"

I was equally impressed by the Inspector's resourceful use of materials at hand. Normally, beavers do not build edible vegetation into their structures. Old dead wood, debarked food sticks resurrected from the pond bottom, muck, fallen leaves, and even human litter (bottles and plastic bags) are materials of choice. Now, however, time was of the essence. The pond had to be saved, and the long-stemmed, rubbery lily plants were pliable and could easily be packed into the interstices between the rocks. Moreover, they were readily at hand, and the Inspector General was working at a fanatical pace—diving, plucking, packing, diving, plucking, packing, diving, plucking, packing. Almost as soon as he disappeared underwater, he bobbed up again and swam for more aquatic vegetation. Nothing distracted him, neither our proximity nor the roar of water rolling over the top of the wall.

Soon we spotted three more beavers steaming toward the broken dam and recognized the two yearlings, Huckleberry and Buttercup, trailing in the wake of Blossom, who was still a few weeks shy of his second birthday. Within seconds, they were on-site, swimming back and forth and taking in information with all their senses. In short order, they followed the Inspector General's lead and went to work pulling up whole lily plants—stems, pads, blossoms, and even roots—and clutching these vegetative masses between their short arms and chin, they too disappeared underwater, where they remained for however long it took them to locate and plug up an underwater chink in the stone wall.

I had never before observed beavers behave like the busy workaholics they are reputed to be, but on this night they seemed to understand that their very existences depended on nonstop action. Trip after trip they made to shrinking lily patches to obtain more caulking material. Down and up the four of them bobbed, like a crew of scuba divers on a rescue mission. Watching them, I began to hope that what they were doing might actually save the pond, that the soft vegetation they were using might actually hold back a deluge.

Still the backside of the wall continued to spurt great volumes of water. And over the wall's irregular rim the pond continued to empty itself. And all that water converged on the far side to form a stream that gouged an ever deepening channel in its seaward rush.

So this is the awesome power of water, I thought to myself. Small wonder we harness it to our service. But who would dream that a quiet beaver-pond contained such energy?

I despaired that such a force could be checked. But the beavers continued to battle it with lilies, one handful at a time. Was there not some significance in this? And did not their perseverance imply that success was at least possible? After all, wild animals do not spend their precious energy reserves indiscriminately, with no real potential for return. Natural selection eliminates those individuals who squander hard-won calories pointlessly. So it is that wolves test many possible victims to determine how vigorous they are before running down the one they stand a chance of catching. So it is that wild cats make short, fast dashes at prey, but abandon the effort if they fail to catch their dinner at once. Would not beavers too possess a sense of what can and cannot be attained? Would not beavers too know when to quit?

Yet these four beavers continued to hustle. Once again the Inspector General made an attempt to buttress the downstream face of the unstable wall with hard, woody material. Once again he swam to the far shore and returned with an oversized laurel cutting, which he tipped over the crest and anchored between the rocks. When this held, he returned for more and was soon joined in this endeavor by two of the younger beavers.

I was struck by how well the animals worked together. Although their labors could not be described as coordinated (each beaver performed tasks according to his own inner dictates), nevertheless they did not seem to get in each other's way. Nor did any one of them show signs of annoyance when his or her work was undone or redone by another. The Inspector General, in particular, rearranged sticks placed by the others. Was this some kind of compulsive busywork on his part, or was he actually making improvements?

I recalled an experiment performed by two Swiss researchers, A. Aeschbacher and George Pilleri. Noting that lodge-building beavers frequently rearrange sticks that have already been pinned in place, the two men marked forty branches and returned a month later to see which of these had been moved and where. What they discovered was that beavers sort lumber according to size, switching pieces around to meet specific building requirements. Long sticks, for example, had been removed from the lodge's low-vaulted entryway and worked into the high-vaulted main dome of the structure; by the same token, short sticks had been removed from the high-vaulted main structure and put to better use in the low-vaulted entryway.

Whether or not the Inspector General was doing something equally practical, his efforts did seem to tighten the leaky structure. By contrast, Huckleberry's revisions were inadvertent. When inserting a contribution he had brought, he often dislodged work already done by others, thus causing the sound of escaping water to augment sharply.

As the hours passed, I began to wonder when the four beavers would take time out to eat. It seemed extraordinary to me that they had not already done so. In two years of watching the colony, on no evening had I failed to make the notation: "Beavers spent the

first hour feeding." Like all plant eaters, *Castor canadensis* must de-
vote a large part of its waking life to ingesting food. Yet on this
night the work crew seemed perfectly able to ignore that need.
Only Huckleberry took a brief lunch-break to snack on a glob of
lilies the Inspector General had packed onto the work-in-progress.

I wondered about Lotus and Lily. Where were they? Was it
necessary for *two* beavers to remain in the lodge with the kits now
that all its entryways were open to the world?

When darkness obscured the identities of the four individuals we
were watching, we stayed on, straining to catch sight of their rotund
shapes moving about in the water. Even after these vague forms
could no longer be discerned, we lingered awhile longer to listen
to the sound of diving. In the dark, the soft plopping of beaver
bodies slipping underwater was as evocative as the dip-dip-dip of
a canoe paddle. Just before giving up for the night, I shined my
light along the curved jetty and discovered the Inspector General
on its downstream side, hard at work plugging it from behind. This
was a first observation for me. Until then, I had not questioned a
widely-held belief that beavers do not work from the backside of
a dam.

Driving back to the cabin, I tried to elicit words from John to
the effect that the beavers could and would save their pond, but
without success, and his refusal to patronize me with false en-
couragement created tension between us. The most he would con-
cede was that the animals' Herculean try was cause for wonder.

That night I slept poorly, haunted by the events of the day and
night, and in the morning I prepared myself for the worst. Perhaps
the stone wall had collapsed while we slept, releasing a surge of
water that washed away all the work the beavers had done, together
with what remained of the original dam. Perhaps, upon arriving,
we would find a mud flat where Lily Pond had been.

As we approached the point on the path where all would be
revealed, I lagged behind, keeping my eye on John's back for some
warning of what lay ahead. For a moment after reaching the top
of the hill, he kept me in suspense. Then he swung around and
faced me with a wide grin.

"It's held!" he shouted.

"It's held?"

I raced to his side, and on that high vantage point we hugged each other and cheered the beavers—who by now had retired to their lodge for a well-earned rest. The pond was low, but the funny little connecting piece John had laid out and the beavers had buttressed and chinked had stood through the night and was doing a job. We ran downhill to inspect it. Though it was still leaking a good deal of water, the beavers had greatly strengthened it during our absence, thus reducing outflow by perhaps two thirds.

To say that that scrappy-looking insert appeared out of keeping with the original dam would be an understatement worthy of inclusion in the *Guiness Book of World Records.* Not only did it bulge into the pond in a most ungraceful and incongruous manner, but its cover of fresh vegetation—leafy cuttings, pink and white blossoms, lily pads, and black roots—called to mind a homecoming float that didn't make it to the parade. By contrast, what remained of the original dam looked a sensible affair. Its neat backing of dead, bleached branches, all set at the same vertical angle against its downstream face, resembled a tightly made picket fence. And its well trodden crest of hard-packed earth was almost indistinguishable from many other sections of the pond's natural shoreline.

"This thing you and the beavers have put together looks like what it is—a dam dreamed up by a committee," I told John. Then I gave him a big hug for the part he, as committee chairman, had played in designing it.

Of course the dam insert was a fragile affair and needed a great deal of reinforcing. So that morning we enlisted the help of three small boys who had come to the pond to fish, and the five of us toted load after load of badly needed sticks to the dam site and dumped these into the water. A good deal of what we brought was material from the original dam that had been washed one hundred and fifty yards downhill before getting hung up on stumps and brush.

"This represents years of beaver cutting," John remarked, as he trudged uphill bearing a cumbersome load on his shoulder. "The colony couldn't make up for such a loss of material if they were to do nothing but fell trees for a month."

We worked all morning, recovering what seemed to be several cords of thick sticks. But John remained convinced that more reinforcement would be needed and he worked for several more hours backing the jerry-built wall with rocks.

When at last he felt satisfied that the still leaky structure was somewhat more stable, we followed the example of our nocturnal beaver friends and returned to my cabin for a fast nap; for it was our intention to spend most of that midsummer's night at the pond, watching what the beavers would do next.

Chapter Twenty-two

J ohn and I were back at the dam before the beavers woke up.
"I bet they sleep late after last night's workout," John said.
But at 6:05 a beaver emerged from the lodge.

"What an extraordinary thing he's doing," I said, peering at him
through my binoculars. "He's pulling a big log off of his house."

John raised his binoculars and affirmed that indeed the big beaver
did appear to be doing something like that.

"I think it's the Inspector General," he added.

We watched as he hauled a six-foot-long bole into the water and
began towing it from lodge to dam, a distance of one hundred yards.

"Could he be removing material from his lodge to use in his
dam?" I asked in amazement.

The implications of this act made my head reel. Was it possible
that the beaver had awakened with such clear recall of the previous
night's events that he *foresaw* the need to bring lumber to the dam
site? It is one thing for a beaver to collect damming material after
being confronted with the sound and sight and tactile sensation of
an actual leak. It is quite another for a beaver to make advance
preparations to meet a calamity he simply recollects has taken place.
For from such a distance the Inspector General could not possibly
see any break in the structure, nor could he *feel* or *hear* water running
out of the pond, particularly since he was upwind of the dam. This
then was no automatic response—conditioned or inherent—to the
presentation of visual, tactile, or auditory stimuli. Some higher
psychic function seemed to be involved. Equally noteworthy was
his readiness to remove a log from the lodge roof, an act that must

have aroused some conflict in him, given that most animals treat their "nest" as inviolate and would not dismantle even a small part of it to advance another project. It would appear, then, that the memory of the previous night's emergency had been clearly retained by the Inspector General, even after passing a day in sleep, and the impulse to meet that challenge again, this time with the proper materials in hand, must have taken precedence even over the animal's strong attachment to his place of shelter.

I was privileged to witness more thought-provoking behavior when the big patriarch arrived at the dam with that roof beam in tow. The water there contained a tangle of floating wood, the stuff that John and I had brought and dumped while the beavers slept, and the Inspector General's reaction to this windfall was to drop his payload and begin swimming this way and that through the log jam, sniffing one old branch after another and expressing his beaverish excitement in loud tones.

"I hesitate to say this, but that animal sounds like he's rejoicing," John whispered.

I could hardly suppress my glee.

"Maybe 'marveling' would be a better word to describe what we're hearing."

Whatever the beaver's subjective feelings, which were beyond our power to ascertain, they prevented him from settling down to the work at hand. He swirled about through the raft of branches like an excited dog who has discovered a "delicious" smell to roll in, and he never stopped "talking" about his find.

"It's been worth all our efforts just to see this," I said.

Then I gave in to a desire I usually try to suppress and spoke to the wild animal. Immediately, the beaver faced me, and seeming to take in my words, responded with a few soft "uh-uh-uhs" of his own. And so we carried on a dialogue lasting perhaps a full minute, he in his tongue and I in mine. Afterward I believe we both felt better for having made that attempt to communicate across the species barrier.

In due course, the beaver set to work. After prying one stick from the tangle of floating lumber, he crawled across the debris, mounted the crest of the dam, and toppled it onto the rock reinforcement that John had built against the structure's back side. This done, he repeated the process again and again, systematically dis-

Time is of the essence. The Inspector General removes a log from his lodge and tows it one hundred yards to the broken dam, where repair materials are desperately needed.

entangling and toppling one stick at a time. His motions lacked the frenzy of the previous night. Now he worked steadily and without haste, at what I knew to be *Castor canadensis*' normal pace. Clearly, the emergency was over. Now the grunt-work was proceeding.

Soon we spotted four more beavers en route to the dam, and all but one were towing logs they had lifted from the lodge.

"That is the darnedest thing I ever saw," John said. "These beavers plan their work better than my students, who can't even think to bring pencils to class." (At the time, John was teaching a course in nature writing.)

Like the Inspector General, the four beavers, on arriving at the dam, commented aloud over all the floating lumber there. Blossom and Lotus gave vent to their excitement by cavorting about, leap-frogging over each other's backs. Buttercup, after swimming in circles and vocalizing, suddenly spotted us and slapped the water. But when this failed to drive us away, she took no further notice of our presence and set to work on the dam. And Huckleberry, eager worker that he was, pushed through the mass of floating logs and sticks to nip a top branch from the dam's flowery crest, and clutching this bouquet of white blossoms in his teeth, he scrambled

Above: To strengthen their rickety repair job, the beavers make use of the washed-out dam lumber John and I retrieved during the day. Below: A yearling beaver appears pleased to discover the building material John and I dumped by the dam.

188

down the steeply sloping backside and rammed it into what must have been a wrong place, for a new hole opened, through which water spouted.

Beavers back their dams with thick sticks, which not only support the structure, but snag a good deal of debris that would otherwise wash over or through it. Thus to some extent a beaver dam is self-sealing. By contrast, the animals pack mud onto the frontside, the side that is underwater and out of sight. Normally, mud is readily available just at this site.* Now, however, much of that precious resource had been washed away during the torrential outpouring caused by the break. To plug and reinforce the frontside, therefore, the beavers improvised. In addition to using uprooted lilies, they packed it with grass, small brush, and tourist litter (bottles and plastic wrappings), most of which they obtained on land.

To say that we enjoyed watching the animals repair their fragile patchwork dam on that longest day of the year might suggest we lacked empathy for the hard-working beavers. The fact was, however, they no longer aroused our concern. They seemed not a whit agitated; on the contrary, they performed their various tasks as calmly as a crew of professionals who know precisely what must be done. Moreover, I was glad for this rare opportunity to observe building behavior in good light, hour after hour, and at close range. So intense was their absorption in the work at hand that they did not spook even when I moved about in their midst to photograph them.

I soon saw that every animal worked on both sides of the dam and that sometimes an individual cut a branch to size before jiggling and wedging it into a particular space. Most repositioning of sticks was done by the Inspector General, who did justice to the name I had given him. He gave the impression of an overseer whose standards can never quite be met, for he persistently redid work performed by the others.

To my surprise, I also observed several instances of cooperation between animals, behavior I had not seen before. Twice I saw two beavers work as a team to drag an oversized log across the jam of

---

*Topsoil, washed from the land, is carried on a seaward course by streams and rivers, and only at places where the current is checked—such as beaver dams—does the water become still and let go of its precious cargo, thus allowing life-sustaining loam to settle to the bottom.

Above and on page 191: Beavers work nonstop, on both sides of the dam—packing mud and vegetation into rock-wall crevices, disentangling lumber, cutting sticks to size.

woody debris, then tip it over the crest. In another such joint effort, I watched Blossom elicit help from Lotus, after he had made several unsuccessful attempts to extricate a single branch from the tangle of floating material. Responding to her sibling's grunts, Lotus took hold of the interlocked branches with her teeth and "hands" and steadied the mass, while Blossom pulled out the twiggy limb he wanted. Later in the evening I saw this same piece of business repeated.

I also noticed that the Inspector General followed up on work poorly performed by Huckleberry. After the yearling fixed his branch so that one end was loosely pinned under several others, the big male crawled over and gave that stick another mighty shove, thus securing it. At one point he improved working conditions for himself (and incidentally for all the beavers) by opening up an access route through the floating snaggle of wood that fronted the dam, cutting and pulling away a large section of it.

All that night the five beavers worked, and only Lily failed to show up. By morning the appearance of the dam was greatly altered. With the addition of the dead wood we had brought to the site, it no longer looked like a flowering hedgerow. John's backing of rocks had been neatly covered with vertically placed sticks, including one particularly long and heavy log we hadn't expected would be used. Early in the evening we had watched various beavers attempt to do something with this humongous bole, only to give up. In the end, two or more animals must have pulled and lifted together to hoist it over the crest.

All in all, the repair job was a success, though a precarious one. Some leakage continued to flow through and under the dam, but hardly more than normally escapes such a large impoundment. Still, the structure needed many more loads of mud and sticks to strengthen it. Moreover, the pond had fallen to a very low level and a great deal of water would have to be captured over the next few weeks. The experience of the Skipper and Second Mate clearly demonstrated the danger posed to beavers of wintering in a too-shallow pond. To prevent another such near-calamity, the dam would have to be raised again and again so as to contain every drop of rainfall and runoff that fell over summer and fall.

That afternoon I said goodbye to John, the hero of the weekend, and prepared to settle back into my routine of watching unmolested beavers be beavers. I hoped to be present for the debut of the kits from the lodge, due to occur at any time. And I was curious to see what, if anything, the colony would do about their exposed doorways.

But even as I drove to the pond, the colony faced a new peril, and by the time I came to the vantage point along the path, the crisis had become full blown. What I saw from the top of the hill was a terrible commotion at the lodge. Water was spraying into the air. It took me a moment to make out the cause of all the turbulence: two swimmers wearing fins on their feet and clenching snorkles in their teeth. Then I spotted a third person, a woman standing on the dam one hundred yards from the swimmers, and she was throwing rocks at the Inspector General and shouting at him to get away.

I raced downhill shrieking at her to stop pelting the beaver, who for some reason did not dive underwater or swim away, but maintained his position and looked straight at her. I all but yanked her off the dam.

"What in the name of heaven do you think you are doing?" I demanded to know.

She was unflappable.

"I don't know what you're so excited about," she answered. "I'm just defending myself from that animal."

I could only gasp at this outrageous statement.

"He's going to attack me because my friends are trying to get inside his lodge," she went on.

I don't like to remember myself in the kind of rage her words produced in me. To be so filled with fury is quite unlovely. I bellowed at the swimmers to get away from the beaver lodge. I *ordered* them to shore. I threatened to have them arrested. I shouted language at them I didn't know I knew.

In hindsight, I suspect the only reason they responded to my outcry was because I happened to be wearing a khaki army jacket, and they mistook it for a park uniform. Even so, they moved away from the lodge with all deliberate slowness, and when they rose out of the water in front of me, their manner was defiant. They demanded to know who the hell I thought I was to tell them what they could and could not do.

At such times I realize that I am as capable of violence as the next person. Fortunately, I learned early in life that it doesn't pay to act on that impulse. So I lashed out with words, castigating the two men, letting them know the stress their actions had created in the animals, informing them that the law was on my side, spelling out the penalties imposed on anyone who harasses a beaver colony or damages their works.

After my initial outburst, I calmed down a bit and made an attempt to educate the troublemakers on the effect their intrusion likely had on beavers who were guarding young. I told them about the vandalized dam and described the tremendous effort the animals had made to repair it. Then I pointed out how low the water had fallen and called attention to the open entryways to the lodge, a condition bound to be stressful to the animals.

The men were unmoved.

"You must think you own this place, the way you come on," one of them sniped.

And the woman piped up, "Park officials don't want beavers here anyway."*

---

*When I reported this incident to game warden Kenneth Didion, he suggested the party may have been trying to steal kits to sell.

Despite their resistance to everything I said, I sensed that they were anxious to leave and so I retired to my cove to wait and watch for any beavers that might show up on the pond. In time I heard a car start up and drive away and knew that this latest menace to the colony's well-being had departed. Not, however, before they slipped a nasty note under my windshield wiper.

Meanwhile, no beaver put in an appearance. Finally, at nine o'clock, the Inspector General bobbed up and began patrolling the water in front of the lodge. I had never seen him so agitated. Back and forth he swam, slapping his tail repeatedly. Again and again he started for the dam, but lost his nerve and turned back. Once he headed out with building material in tow, then dropped his load and turned back.

It was nearly ten o'clock before a second beaver emerged and approached the cove. I saw it was Blossom, the most trusting member of the colony. But now even he stopped short of his destination and eyed me with suspicion. For several seconds he tested the air with his nose, as if trying to make up his mind whether or not to proceed. Finally, the impulse to retreat triumphed, and he headed away. I tracked him with my binoculars until he swam into a patch of lilies and became lost from view.

It now struck me how important had been that solid mat of aquatic vegetation, which two years ago had completely blanketed the pond and concealed the beavers from view. When I first came upon the place the animals had been so difficult to see I had trouble convincing myself any existed. Now so much aquatic cover had been uprooted and eaten that the beavers had become conspicuous—a situation that may have contributed to the sudden spate of attacks on them.

I left at midnight, when I could no longer see, and returned the next morning to find a six-inch lip had been added to the dam during the night. This work had been done without the prompting of any noisy outflow and left me wondering what had motivated the beavers to take such a farsighted step; for by adding that much extra height to the crest, they were now assured a rise in water levels with every summer shower. In addition, all remaining leaks on the frontside had been plugged. The dam was tight.

That night I was relieved to find myself the only visitor to the pond, but I saw little of the beavers. Normally, all but the sitter-

The backside of the repaired dam. Only an insignificant amount of water now escapes.

on-duty emerged around six. Now I waited until nine before a lone animal straggled out and headed for the marsh at the inlet end of the pond. There, concealed by tall saw grass, five beavers eventually congregated.

I was surprised that they felt no compelling need to examine the beautifully repaired dam. Did they *remember* that it was now watertight? Not until dark did I catch sight of a beaver silhouette headed in that direction. It was the big male on his way to make his nightly inspection. I eased down the shore to spy on him, but was quickly discovered. Tall geysers shot into the air as he spanked the water again and again. And when I stood quietly and refused to leave, he did.

So you have become leery of me again, I mused.

While this reaction would certainly make future observations more difficult, I was nevertheless relieved to know that his tolerance of my ubiquitous presence had been so tenuous. I hoped the same would prove true for the others. Otherwise, with cover shrinking and park visitation on the increase, more trouble lay ahead for the colony.

Still, it made me sad to think that wild creatures can never let down their guard against human beings.

195

A high lip has been added to the finished dam. This will help contain rainfall and runoff and thus assure that pond levels steadily rise.

Chapter Twenty-three

Exactly one week after the dam was vandalized, two kits emerged from the lodge. They were escorted by two adults and a yearling, upon whose various backs they rode. I obtained only glimpses of them from my station across the pond before they returned to the house.

So they had made it, open door or no! I waited and watched for an hour but didn't see them again, although a pair of muskrats who made their home in the wall of the beaver lodge and who were roughly the same size as the babies managed to confuse me by putting in sporadic appearances. To distinguish beaver kits from such look-alikes requires close scrutiny. I decided to make my way to the other side of the pond, where no eighty-yard water gap stood between me and the lodge. This meant I would have to walk gingerly across the newly repaired dam, then crawl one hundred yards through dense laurel on my hands and knees.

The fresh material in the dam had not had time to settle and harden, nor had it been tamped down by the many animals who in time would use the structure as a foot bridge. Here and there, however, the beavers had hoisted a good-sized rock from the bottom and placed it on the mushy topping. Slowly and with extreme caution, so as not to damage the structure, I stepped from one of these to the next.

It is precarious business crossing a beaver dam. Should you begin to totter, you must quickly make up your mind which way to drop. If you tilt downstream, you probably will remain dry, but you stand a fair chance of being impaled on any of the hundreds of vertically placed, pointy sticks the beavers have cut and used to reinforce the back of the structure. If you lean the other way, likely you will find yourself over your depth in mucky water, for here the beavers repeatedly mine the bottom for mud to pack against the dam's underwater frontside. The safest time to cross is, of course, in winter, when the crest has become frozen. And the safest dam to trod upon is one that has been in place long enough for plants to have become deeply rooted in it.

I did not, however, have the luxury of choosing my time or place of crossing, and so when I made it to the far side of the Lily Pond impoundment I experienced some relief. From there I spotted Lily, heading my way, and decided to wait and find out what she was up to, for I had seen little of her during the past few weeks.

At first I failed to notice Huckleberry, swimming like a whale calf close at her side, but as the two approached, his antics caught my attention. From time to time, he touched her shoulder with his nose, then tried to climb on her back. This was odd behavior for a yearling beaver, and Lily did not seem to go along with it. Each time he attempted to ride her, she dove and dunked him. This rejection did not, however, put off the pesky youngster. He continued to plague her. What was going on? Had the presence of new kits in the lodge caused this one-year-old to regress to a more infantile stage? I had seen yearling mustangs act up when their mothers returned to the harem with new foals at their heels. Some of these one-year-olds would try to crowd their younger siblings off their mothers' teats. Was it possible that Huckleberry was going through some such phase?

Whatever was on his beaver mind, Lily did not suffer his behavior gladly. Shaking loose of her pesky offspring, she swam in a wide arc, turned sharply, and made a headlong rush directly at him so that, when their heads met, he was lifted up and out of the water and flipped over onto his back. That brought the adolescent back to reality in a hurry, and he sped off to the dam and began working on it in earnest.

For several minutes before beginning my long crawl to the lodge I remained near the dam to see who else might show up. It was interesting to view the pond from this angle. My station looked so far away. While I was taking in the unaccustomed perspective, Blossom and Lotus appeared. They were in high spirits, porpoised under and over each other, nuzzled each other's sides, and then, to my delight, hauled out of the water and performed a push match on land. Their interest in one another seemed to be growing more intense. By now they were of an age when most beavers depart their parental pond, and every night I counted heads to see if they had done so. My feelings were mixed on the matter. Should they leave, I would certainly miss them; yet I was more than a little fearful they would remain at Lily Pond an extra year. If so, eight beavers would winter together in the lodge, and I couldn't imagine how that many animals would feed themselves now that the pond was so depleted of lily roots.

When Blossom and Lotus swam off, I could think of no more excuses to postpone threading my way through the laurel, an ordeal made the more agonizing by knowledge that I would have to make a return trip through the same tangle. When at last I emerged at my destination, an outcropping along the shore just twenty feet from the beaver house, my face and arms were scratched, my back seemed permanently bent over and I felt on the verge of a heat stroke. Still, it was reassuring to contemplate the odds of anyone else approaching the beaver house by this route. Though generally speaking an island lodge is less accessible than one built against a bank, this particular bank lodge might as well have been completely surrounded by water, so impenetrable was the land route to it. I marveled at how the colony had picked such a protected spot on which to build.

Having completed the two-part obstacle course, my reward was quick in coming. Plaintive cries of infant kits emanated from the lodge, and the water in front of me began to rock gently, indicating that a beaver was at that moment descending the exit tunnel and soon would emerge. I held my breath and watched as the blond yearling, Buttercup, rose to the surface and shook herself. After floating quietly for a few seconds, she began to scout the area around the family residence and sniff the air, but she failed to detect my downwind presence. Would the kits follow their sitter's lead and come out? Or were they old enough now to be left unattended inside the lodge? Perhaps Buttercup had left them only briefly, no

Blossom and Lotus, overdue to emigrate, become inseparable.

longer than it would take her to defecate (beavers prefer to perform this act in water). But no, once out she busied herself in the immediate area. First she dredged up a debarked stick from the bottom, which she inserted over one of the lodge entryways so that it blocked passage. (Despite a rise in the level of the pond, the three exit/entryways still peeped above the waterline.) This done, she repeated the process, positioning several more branches in the same manner. Then she went to work underwater, chiseling and digging out mud and sticks from the lower part of the opening to enlarge it. Thus she succeeded in lowering the doorway by five inches, and when she finished, it was again covered by water.

Was this repair job more evidence of *Castor canadensis*' intelligence? It hardly seemed probable that nature would endow the beaver with an encoded program for dealing with a task as specific as the redesign of an entryway. What an enormous brain the creature would have to possess were that organ wired to meet every possible engineering contingency that might arise. Still, simple units of innate behavior might be combined in a variety of ways to meet a wide range of unexpected situations. Even so, a "foreman" would have to exist in the beaver's head to select and order their proper execution. As I saw it, beavers show too much ingenuity in the many ways they respond to their environment to be dismissed as mere robots. Moreover, there remained the question of timing. The doorway had been exposed for longer than a week. If Buttercup were reacting automatically to a particular stimulus (say the sloshing of water in the exposed entryway), why had it taken her so long to do so? And why had none of the other five able-bodied beavers, who shared her digs, experienced a similar compulsion to engage in this activity?

I am not the only student of the beaver who credits the animal with high intelligence and a capacity for problem solving. The French researcher P. Bernard Richard shares this view. At the World Symposium on Beavers, held in Helsinki in 1982, Richard reported many examples of the animal's extraordinary capacity for adaptation, the signature of real intelligence. In making his case, he described having watched a beaver stack debris beneath a tree, the base of which had been covered by a wire netting to protect it from being cut. When the mound was high enough, the animal climbed upon it and severed the trunk at a point above the mesh.

In another experiment, a piece of bread was suspended by a string and presented to a rat (often cited as an intelligent animal), a muskrat and a beaver. The muskrat and the rat jumped at the bait and attempted to tear and snatch pieces of it. (Try this for yourself; it's like bobbing for apples.) The beaver, by contrast, studied the situation, then cut the string. Richard asserts that this capacity for adaptation, sometimes called "behavior of detour," is well known in the chimpanzee but has been overlooked in the beaver.

There is little question in my mind that beavers work out ingenious solutions to problems. When a single branch of a tree they are dragging becomes entangled in surrounding brush, they do not waste energy yanking on it, but discover which limb is causing the trouble and selectively clip it. If this behavior were nothing more than an encoded response to a particular stimulus (say, the resistance of the object being towed), one would expect to see a spate of random limb-nipping released which might not be satisfied until every branch of the tree had been severed. I never saw anything like that happen.

While still marveling over Buttercup's clever remodeling of the entryway, I caught sight of Lotus approaching the lodge with a mouthful of green grass for the babies to eat or lie on. She traveled by way of the donut ring and thus discovered me hidden in the shoreline vegetation. Slap! Dive! Slap! Dive! Gone! Only a trail of bubbles told me she had located the newly lowered entrance and followed Buttercup into the lodge.

Shortly thereafter the Inspector General cruised past. He, too, carried grass in his mouth, giving him the look of a green-whiskered walrus. He, too, appeared startled at finding me on the wrong side of the pond. But instead of slapping the water and fleeing, he approached to within three feet of me, head elevated and nostrils flaring, and thoroughly checked me out. Once satisfied that I was just one of the pond regulars, like the deer and the raccoon and the otters and the muskrats he often encountered, his state-of-alert posture wilted and he sailed off to complete his delivery.

Now there were three adults in the lodge, two of whom may have been sufficiently aroused by the sight of me on their side of the pond to transmit "lie low" signals to the kits. As time passed, it seemed to me that something like that must have happened, for I no longer heard tiny, importuning voices behind the thick walls.

Finally, the Inspector General emerged alone and headed toward the grassy bank from whence he had come. As he sped past he gave me a long hard look, but did not slap the water. Then Lotus appeared. Her behavior demonstrated that she well recalled not only my presence, but the precise location where she had spied me, for she gave that part of the shore a wide berth as she headed toward the dam. Now only Buttercup remained with the kits, and since to my knowledge she had no inkling that a human spectator was seated just twenty feet from the lodge, I clung to the hope that she would bring them out.

And indeed she did. Within minutes I caught sight of her swimming back and forth in front of the lodge. And clinging to her right ear was a dark kit. And riding her tail was another, slightly lighter one. Together the three of them plunged and surfaced like dolphins, an activity that must have given the babies a good physical workout and some training in diving as well, for kits this young have to work hard at staying underwater. Nevertheless, underwater is where they are inclined to spend a good deal of time. Once allowed out in the pond, they cavort beneath the surface for such extended periods that I often concluded, wrongly, that they had returned to the lodge. Then suddenly the two would pop up in unison, suggesting they had been engaged in some kind of underwater precision act. How I longed to enter their watery realm to watch them.

I don't know what finally tipped off Buttercup to my presence, but all at once she assumed the "alert" posture, sent the two obedient kits into the lodge, and headed straight for me. Three feet away she stopped, idled, and stared, as if she recognized me and was puzzled to see me on that side of the pond.

"Okay, you caught me," I said to her. "No need to make me feel so guilty. I just need to get a handle on the new arrivals. Besides, you don't think I enjoyed crawling over here, do you?"

It doesn't matter that beavers don't understand words. They do detect some quality in the human voice that either reassures them or makes them feel uneasy. To allow them to make such a determination, it is necessary to utter words, however silly they may sound to yourself and to any other human being within earshot. Buttercup listened quietly, let out a huff, and swam back to the lodge. I interpreted this to mean that she was only moderately perturbed, for were she truly agitated she would have hissed or slapped her tail at me. Then she dove into the entryway to join the kits.

I visited the far shore only a few more times that season, for soon the little ones were all over the pond in the company of whatever beaver happened to be on hand to escort them. At first the adults and yearlings vied for the privilege. If one set off with the kits in tow, the little entourage would soon grow as others joined it. At times the kits were so surrounded by family members that I wasn't able to see them. On those occasions, much water sport took place and did not die down until the beaver convoy reached some still-intact patch of lilies and began to feed.

On one of the last evenings that I ventured through the laurel to observe family life, Buttercup was again on duty and, as was her custom, patrolling the waters immediately outside the lodge. This time she promptly discovered my presence along the shore and lost no time before slapping the water with her tail, perhaps as a warning

Buttercup appears perplexed at seeing me near the natal lodge.

to the kits not to come out. If so, they failed to heed it. Within minutes they emerged, but only for as many seconds as it took Buttercup to herd them back inside again.

What close-up views I had of the youngsters revealed they were not as "cute" as had been the previous year's litter. Whereas Huckleberry's and Buttercup's baby faces had been wide and chipmunklike, those of the newcomers were narrow and somewhat mouselike in shape. Well, if I lacked impartiality, their relatives certainly did not. They doted on the infants. To compensate for my biased view, I gave the babies beautiful names. The darker one I called Dogwood, the paler one, I named Daisy.

Their most faithful nanny turned out to be Buttercup, who also happened to be the most beautiful beaver on the pond. Long after the other adults had become quite casual about the youngsters' whereabouts, this yearling dutifully conducted them around the donut ring. Once, however, I did think I saw her give Daisy the slip. For some time she and that lighter kit had been feeding in a patch of lilies not far from the lodge. Suddenly Buttercup departed at top speed, heading toward the inlet end of the pond. In following her movements through my binoculars I noted that she was by herself. Yet when I panned back to the lily patch, I saw that the kit was no longer there. Assuming the youngster had swum underwater to the lodge—for that is where kits normally head when they find they have been ditched—I turned my attention back to Buttercup and tracked her course until she reached the far marsh and tipped up to feed. One moment later a tiny beaver shape popped up beside her. It was Daisy. She had been accompanying Buttercup all the while, swimming underwater for a distance of one hundred and thirty yards.

That seemed quite a haul for a beaver whose lungs were only seven or eight weeks old. Small wonder the colony had become so cavalier about the comings and goings of the athletic youngsters. Their babies no longer required solicitous care; thus they had turned their attention to other facets of beaver life. And something told me it was time for me to follow their lead and do likewise.

Chapter Twenty-four

L otus and Blossom wintered over for a third year at Lily Pond. I was not pleased to see the two-year-olds stay on, for I doubted enough lily roots existed in the pond bottom to feed eight beavers between December and March. Throughout spring, summer and fall, the colony had been most profligate with this important winter resource, dredging it up and consuming it as if it were inexhaustible. And why should they expect otherwise? Certainly beavers don't inventory the roots that remain buried in their pond muck. So how would they know that their horn of plenty was running out?

Nevertheless, that November something prompted the beavers to cache an adequate supply of winter feed for the first time since I began watching them. They cut birch, swamp maple, sassafras and ironwood trees that grew around Square Pond and along the edges of several smaller feeding pools the colony had created above that one. And they sectioned, dragged and floated this harvest over a half-dozen dams, through a marsh of saw grass and yet another two-hundred yards to the Lily Pond lodge. There, after plunging each delivery into a basin they had dug, they headed back upstream, like a fleet of empty tugboats traveling empty to pick up more cargo.

While all this was going on at Lily Pond, the New Pond beavers, as described earlier, had moved to Top Pond, where they, too, demonstrated that their ability to prepare for hard times ahead had not atrophied from lack of use. As a result, I went into winter with some assurance that my subjects were well provisioned. And indeed

Beaver about to section a felled tree.

they were. In late March, when the ice went out and I got a first full count of both colonies, I was delighted to find that all twelve animals had survived the season.

Spring was another story. The beavers were hardly liberated from their ice prison when it began to rain. On April 4, following days of steady drizzle, the chronic bad weather became acute. Six inches of water fell during a twenty-four-hour period, and that was not the end of the deluge. Rain continued to pelt the eastern seaboard for several more days until the ground became so saturated it could accept no more moisture, and rivers and streams overflowed their banks, and little towns were forced to evacuate, and a New York State Thruway bridge gave way, taking with it a car and driver. Still it rained. By then all the dams that shored up the seven pools leading uphill from Square Pond to Top Pond were concealed under tumbling waterfalls. And water climbed the tepee-shaped lodge that sat in the middle of Top Pond until that structure was half submerged and its interior chamber flooded. The Skipper, Second Mate, Sweet Potato and Yam could find no shelter there. What was needed was a break in the weather, so that the Top Pond colony could put a second story onto their house (beavers do this); but rain continued to fall.

Probably the Lily Pond beavers found themselves in much the same predicament, for during this wet spell they abandoned their living quarters and took up residence in a burrow near the marsh. There they tunneled deep into the steep hill that banked the south side of the pond and created a dry living chamber.

But the Skipper and Second Mate and Sweet Potato and Yam enjoyed no such option. No steep bank met the water on any side of their pond. Beavers are not fish. Beavers must have warm, dry living quarters in which to rest and groom and waterproof their fur. Despite the fact that the species spends much of its life in water, any individual who becomes soaked to the skin incurs some risk of dying from pneumonia.

During this stormy period I visited Top Pond every evening, hoping to catch sight of the rained-out animals. Early on I did glimpse a beaver hiding in the culvert that connected Top Pond to the string of pools on the opposite side of the road. But that was the last I saw of any of that family. Either they all moved elsewhere or died of exposure, for conditions were not only wet, but cold. Moreover, as it was early April, few food plants had sprouted around the beaver pools.

It is not easy to locate an unmarked family of beavers that has emigrated. Even radio-collared subjects, if they travel any distance, usually escape detection. Harry Hodgdon noted that sixteen of his forty-four marked colonies disappeared during his two-year study, and only one of these missing families resettled near enough for him to discover their whereabouts. Still, I never gave up hope that I would someday come upon the Skipper and Second Mate and, by some miracle, be able to identify the pair in a new setting. As for Sweet Potato and Yam, they would rapidly grow and change beyond recognition. They were gone out of my life forever.

In due course, it did stop raining. But even then surface water continued to run into the brimming beaver pools, built like a fish ladder up the slope. And groundwater seeped into them also. As a result, they continued to spill over, each one swelling the next to the point of overflowing, and so on down the hill.

As long as the pelting rain had lasted, the Lily Pond beavers had kept pretty much out of sight (paradoxically, this water-loving species does not like to be caught in a downpour). Now, however, they came out and sloshed about and hoisted themselves up and over their discharging dams. Every one of the structures needed to be raised and repaired, and I waited to see where the beavers would begin to work. But to my surprise they ignored the gushing, roaring water. Night after night they visited and fed in their various pools and paid not the slightest attention to the spillage. Was so much noisy flow just too overwhelming? Yet they had been wild to repair the vandalized dam at Lily Pond, and it had spewed a far greater volume than any of these. What then was the reason for their indifference?

After two weeks, enough water had run over the battered beaver dams so that their tattered rims became visible. Amazingly the structures themselves had not collapsed, though all along their crests furrows had eroded, and through these gaps, water continued to escape. Now, I thought, the beavers will surely repair these discrete breaks, and I set up my cameras. But the animals were as unresponsive to these minor leaks as they had been to the unbroken sheets of water that had previously tumbled over the full length of every dam. Days passed. The crest of the Lily Pond dam spewed water at thirteen rupture points, and every night the beavers docked and fed beside these breaks without making the slightest move to plug them. The ponds, after all, were at an all-time high!

Not until the third week in April did the beavers repair their dams. By then the water had subsided so that only the deepest furrows along the dams' crests continued to run and required attention. By waiting, the beavers had spared themselves a good deal of unnecessary work impounding water they did not need or want.

Meanwhile, they had been tending to other business, had placed scent mounds around their new living quarters near the marsh. They also covered the ground above that burrow with logs and brush, thus creating a kind of ersatz lodge there. This wood pile they gradually transformed into a functional dwelling place; for when the brush heap became sufficiently dense, they gnawed their way up into it from their basement warren, hollowed out a living chamber and moved in. At that point the ground burrow became their plunge hole, a tunnel that connected what was now a bank lodge and the pond.

Now that I had only one colony to watch, keeping track of beavers should have been less complicated. Nevertheless, when the eight Lily Pond residents came out of their lodge at night, they scattered across seven pools and left me wondering where to begin. Moreover, with the departure of the Skipper, Second Mate and their young, the Lily Pond beavers had extended their range to include the abandoned Top Pond waterworks. As a result, I was on my feet and moving most of the time, trying to locate and document who was where and when and why.

"It'll get easier when Lotus and Blossom leave," John consoled me.

Those two were past due to depart by at least six months. Still, beaver parents are tolerant of their mature offspring and, contrary to a widely held belief, do not employ aggressive tactics to force late-bloomers out of the family pond. Two-year-olds have been known to remain at their birth pond for an extra year and sometimes longer. Moving out seems to depend entirely on the inclination of the offspring. And of course Blossom and Lotus, having been born out of season, were operating on a screwy timetable. Yet by now they should have experienced an urge to wander. And not just those two. Huckleberry and Buttercup, being only a few months their junior, were also due to set off on their own. I was alert to the possibility that I might lose half my colony overnight.

And so I paid particular attention to those four beavers, watching for signs of restlessness. But as far as I could see they seemed as attached to the Lily Pond waterworks as ever. All marked the area with scent mounds. All fed alongside family members without evidence of strife. Yet one night Lotus and Blossom were gone— had vanished without a trace. I looked for them in all the watery settings that I would find attractive were I a beaver, but they had given me the slip.

Well, perhaps I would do better with Huckleberry and Buttercup. For weeks I kept a sharp eye on them. The fact that they showed no sign of imminent departure was not going to fool me this time. But the two grew fat digging up what lily roots could still be found in the pond bottom. My attention wandered.

Meanwhile Lily was not acting right. I suspected she was pregnant again, but the symptoms I was seeing could not be explained on that account. When swimming, she veered in a clockwise direction and had to work hard to compensate for going off course.

She remained close to the far shore near the new lodge. And on those few occasions when she did come to the north shore, I saw that she was badly in need of grooming.

Beavers are fastidious animals. They groom themselves frequently, more than once a night, for it is essential that an animal who swims through swampy, algae-covered water make efforts to keep clean. To perform this ritual, a beaver hauls out of the pond, sits on his tail and works over his tubby body with all four paws. Every particle of debris and scum is carefully removed, often with a special pincer claw that grows on the animal's hind feet just for this purpose. To scrub his face and comb his belly and hips, places beyond the reach of the hind grooming claw, a beaver uses his forefeet. Watching one of my colony members go over his body with both handlike front paws always made me laugh, for the routine brought to mind a fat lady struggling to pull on a girdle.

The beavers' assiduous attention to the care of their coats serves more than one purpose. Mutual grooming has a soothing effect on participants and strengthens family ties. For the most part, beavers groom one another not with their paws but with their mouths, each nibbling the other's fur with exquisite gentleness. The act looks extraordinarily erotic and appears to me to be incited by the application of castoreum to the fur, for I have watched a beaver comb himself for half an hour without assistance from a companion who was feeding by his side. The moment the groomer rubbed his cloaca and then passed his forepaws across his chest fur, however, his solo act turned into a duet. Immediately his companion rushed forward and began to nibble him up, down and sideways, stopping only long enough to make a fast application of castoreum to his own fur. At that point the two engaged in what looked like a passionate kissing session, each mouthing the other's shoulders, neck and back.

The importance of grooming can be deduced by the amount of time beavers spend at it, as long as an hour at a stretch. Surely, care of coat helps waterproof an animal, for when fur is in disarray, water can seep through and come into contact with the skin. By contrast, when each hair lies in place, the animal's hide is shielded. It is generally believed that beavers gain further protection by smearing castoreum on their fur. I seldom saw a *solo* groomer rub his or her cloaca to obtain this oil, though I often saw water pool on the

Beavers are fastidious groomers. A special tweezer-like hind claw has evolved to help the animal remove any debris that clings to its fur.

Blossom rubs his cloaca to obtain a castor gland secretion, which he then spreads over his coat. Lotus responds in kind, after which the two animals engage in a passionate grooming session.

animal's back and roll right off when he moved. Perhaps a little of the viscous substance goes a long way.

Now Lily's coat was so disheveled it was obvious that neither she nor anyone else had been tending to it. Moreover, a large swollen wood tick had fastened itself inside her ear, a part of the body a beaver normally keeps clean. *Castor canadensis* plays host to a particular beetle, but I had never before seen a tick hanging on any of them.

One night I caught sight of her out of the water, where I was able to study her more closely. Her left front paw was so badly swollen that all its "fingers" were splayed and its palm skin had split, revealing angry, pink flesh. It appeared to be paralyzed, for she carried it aloft and made no use of it, even when she ate. I watched as she single-handedly grasped and twirled a branch she was debarking, no mean feat. To help steady it against her nibbling teeth, she braced the loose end against a boulder.

My heart went out to her. Was this condition a wound or a disease? Did this hand injury explain her inability to swim or even walk in a straight line? Or was there more wrong with her than met the eye?

"Could it have been a trap injury?" John asked when I told him about it.

"I can't tell, but it isn't likely that a poacher would trap for beaver now that the animal is in its summer pelage," I said. "It might be she failed to jump clear of a tree she felled. Beavers are often injured, even killed by the dangerous work they do."

"Maybe she has a fish hook in it, and the wound has become infected," John suggested.

Since the grotesque-looking hand might be symptomatic of a serious disease, I decided to call Kenneth Didion to see what he could tell me about it. But he, too, was mystified.

"We could live-trap her and see if it's a condition we're able to treat," he suggested. "I'll speak to our state biologist about it."

Meanwhile I took a closer look at Lily and saw that she had enlarged nipples. So there were kits in the lodge! That cast a whole different light on the matter. Live-trapping her would threaten their survival, for in addition to being physically impaired, Lily was an aged animal, and the stress of capture could easily kill her. What then would become of her kits? Moreover, even if she survived such an ordeal, she would have to be tranquilized during capture, and how would the chemical used effect her suckling young?

I called Didion back to say I didn't think it a good idea to separate Lily from her infant kits. My point was a moot one. The state biologist he had consulted had declined to interfere with natural processes that regulate beaver populations.* There was nothing to do but wait and see if nature would heal or kill her.

Once again I was delighted by the comings and goings of the kit-oriented beaver family. Buttercup, Huckleberry, Dogwood, Daisy and the Inspector General all brought greens to the lodge. And once again Lily had plenty of help watching over her infants— every age class took a turn at baby-sitting and every change of shift went off without a hitch. At no time did I count more than five beavers out on the pond at once.

Meanwhile, a new lodge was being constructed halfway between the old family homestead, which had rapidly fallen into disrepair, and the beavers' current habitation near the marsh. I could not fathom the purpose of this new dwelling. No beaver made use of it. Like some of man's ill-conceived edifices, it stood empty—a monument to the beavers' need to keep busy. That is, until the day an otter family began poking around the marsh.

I always enjoyed watching otters at the beaverworks. Lily Pond must have been one stop on a much larger fishing circuit covered by a particular female, for she and her young appeared and dis-appeared erratically. During their two- and three-day layovers I was much entertained by their antics. They swam more swiftly than did beavers, cavorting underwater and popping up suddenly where I least expected to see them. They seldom paused, except when an unusual sound or sight or scent aroused their curiosity. Then they would stretch their necks above the surface and rotate their heads from side to side like submarine periscopes scanning

*I can't argue with this philosophy, but I do feel a need to point out that state wildlife managers are as inconsistent in adhering to it as I am. When serving their hunting constituency, they see nothing wrong in promoting surplus numbers of game birds—creating nest sites for ducks and introducing quail and other wild fowl into areas where they do not normally occur. By contrast, they look upon the demise of most nontarget animals as natural population regulation and seldom come to the assistance of an individual in trouble.

for enemy ships. Not infrequently, one would flip over and swim on its back or pick up a floating bottle or some other odd object and play with it. Once I watched a young otter capture a painted turtle he had no intention of harming. He simply held it in his front paws while he did the backstroke around the pond. (How is this different from a child playing with a rubber duck?) On another occasion John and I were astonished to see two daredevil otters make sport of an enormous snapping turtle, repeatedly try to make the predacious creature stand on end. Otters are irrepressible.

Generally, the beavers ignored the presence of these gadabouts. That is until one evening in early July when the mother otter and her two offspring ventured too close to the marsh lodge where the beaver kits were confined. When the three interlopers vanished into a bank hole not six feet from the beavers' entryway, the Inspector General steamed over to investigate. For several minutes he swam back and forth before the crevice, studying it. Then he headed directly back to the natal lodge. That night, as soon as it grew dark, the kits were moved to the log hut that had been standing empty since being built. Thus, what seemed a building gaffe came to serve a purpose. For there it stood, ready and waiting to house vulnerable kits when, quite unexpectedly, the beavers felt an urgent need to move them.

Chapter Twenty-five

L ily was not thriving and she turned to me for help. One evening, after spotting me on my viewing rock, she pulled herself out of the water, wobbled up the bank and looked me straight in the eye. I was taken by surprise, for some two years had passed since I had taken action to discourage the colony from putting that kind of trust in me. Since then what little aspen I had brought to the pond I carefully laid in place in advance of the beavers' emergence from the lodge, so as not to be identified with their find. Then I had retired to a place of concealment from which to view them through binoculars.

But now I saw that the only one deceived by this ploy had been me, for it was clear that Lily was now appealing to me to be her provider. Why else would she suffer the arduous and perhaps painful climb up the bank to confront me? Why else would she risk a face-to-face encounter with an alien species, given her current inability to make a quick getaway? And why else was she at that very moment fixing beseeching eyes on me and addressing me in the importunate tones of an infant kit?

"Uh, uh, uh, uH, uH, UH, UH!"

"Oh Lily, you poor dear, what do you want?" I asked.

She responded with more wheedling sounds.

"I don't have any food for you," I said. "Aren't you able to get your own?"

Though the pond by now was short on lilies, its shorelines and banks had produced a hefty crop of sedge grasses and ferns, and

Lily is ill and appeals to me for help.

behind these grew masses of blueberry shrubs, which offered the beavers the woody twigs they needed to file their teeth on. Such plants represent normal spring and summer browse for a good many colonies, and now that the Lily Pond beavers had discovered the fare, they seemed to be thriving on it.

But suddenly it occurred to me that Lily might experience some difficulty mounting the bank to obtain these foods, for her hip bones jutted out like those of a thirty-five-year-old horse, and her ragged coat could not conceal how rickety her frame had become. Were her nursing kits depriving her of every bit of sustenance she took in? I had seen a stray bitch as emaciated as she and later found her pups to be fat. Perhaps Lily was going through some such stage, one that would pass after the kits were weaned.

But when she turned and staggered downhill I knew otherwise. Her motor control was so poor that she fell and rolled onto her back and lay before me with all four feet sticking straight up in the air. It cost her a good deal of energy to right herself and proceed the rest of the way to the pond, and while she struggled to attain that simple goal, I felt a kind of embarrassment at having seen this wild animal in such an undignified position. Once she reached water and was buoyed by it, however, she seemed to regain mastery over her body.

Throughout this incident, John had been seated some distance uphill from me. Now he appeared at my side.

"She's starving," he said. "Whatever else is wrong with her, she's desperate for food. You better feed her."

"I have every intention of doing just that," I said.

The next night I brought aspen branches to the pond, but did not lay them out for fear the other beavers would discover and consume them before Lily appeared. I settled on this food knowing that it, above all other, is relished by *Castor canadensis*. Moreover, it probably offers more useable nutrition per square inch than any tree growing. That's because *Populus tremuleides* is able to carry on photosynthesis not just with its leaves, but through its bark, too. Aspen and related poplars are unusual in this respect. Most tree species depend entirely on their green leaves to transform the sun's energy into matter, for their bark is pure fiber, a mere covering to protect the thin layer of life-sustaining cambium that lies just beneath it and against the tree's woody core. This layer of cambium contains cells that promote the growth of a tree and is the substance that bark-eating animals seek when they gnaw on the trunks of most trees. But aspen provides more than just cambium to the porcupine and elk and moose and deer and mice and beavers that feed on its trunk and branches. Its bark is as nutritious as a green leaf, for like a green leaf, it captures the sun's energy and converts it into the stuff of life. Not surprisingly, the trunk of a mature aspen has a green tinge.

"Here she comes," John alerted me, and I crept down the bank and laid a sprightly, fluttery-leafed aspen bough on the edge of the water.

Lily swam directly to it and, after nipping off a sprig, bunched the leaves with her good "hand" and fed them into her mouth.

I backed away and climbed onto my rock, from which I had a clear view of the frail beaver. When supported by water, she appeared almost normal, although she experienced some difficulty holding and rotating the branch. Still, her appetite seemed healthy enough.

When my first offering had been consumed, I delivered another.

And when that was gone, I carried still another to the shoreline. Each time I crept downhill to place a bough within inches of her nose Lily remained absolutely quiet, and her grizzled face expressed such trust that I wondered how any human being could harm such an animal. Yet hundreds of thousands of beavers are slain each year for sport and profit, despite the fact that even some of the hard-bitten trappers who commit this act like to tell stories of the animal's winning ways. When a beaver is caught in a land set, they say, it cries real tears. While awaiting the blow of a club about to be lowered on its skull, they say, the beaver covers its head with its forepaws. As far-fetched as these tales sound, they likely are based on actual observations. Scientist L. S. Lavrov, in a report before the World Symposium on Beavers, verified that the North American beaver does indeed produce "a copious emission of tears" when under duress, and is likely to do so when "manually restrained."

I fed Lily on two more occasions, and then she vanished for a few days. During the last of those feedings, all the other beavers discovered what was going on and wanted in on it. There was no way to discourage them from snipping pieces off the boughs I slipped to Lily. Once in her possession, it was her business to defend them, and for the most part she did. When approached by a would-be looter she protected her space by vocal protest. Yet this did not always prevent a sneak thief from making off with a piece of her bough. Though beaver families are amazingly congenial at communal feeds where there is plenty to go around, when food does not reach, individuals resort to subterfuge to gain part or all of another animal's portion.

Now Huckleberry seemed to sense that Lily was no longer her old, powerful self and he became progressively more bold in his attempts to appropriate her aspen bough. As long as she clutched the coveted branch in her teeth, she had no trouble defending it. But when caught holding it with her one useful "hand," she was no match for him. He snatched it from her and made off for water. Lily, however, was not about to be bested by her upstart son. She waddled to the pond, swam after him, and bit him in the rump, causing him to produce a yelping sound I had never before heard from a beaver. Huckleberry dropped his booty, and Lily recovered it. She then returned to the shore, where, for a time at least, she fed in peace.

Huckleberry mounts the bank with theft in mind.

Over the next ten days I did not get another close-up view of Lily, for once her new kits came out on the pond, she spent every evening with them along the far shore. Even at such a distance, I had no trouble distinguishing her from the others, for her skewed swimming pattern set her apart.

"The kits are out," I reported to John. "I guess they'll make it now. Whatever happens to Lily, you can be sure the others will look after those babies."

As with previous litters, the entire colony shepherded the babies about. The moment the two little ones emerged in the evening, every adult within viewing distance converged on them, and much greeting, nose-touching, and splashing ensued. Guided tours to nearby lily patches and piggyback rides were routine activities. And now I noticed a new behavior. Now and again an adult would dive under a kit and bump him from below, lifting him up into the air so that he came down on the water with a splash. Both youngsters appeared to enjoy this sport; at least they swam back for more. The whole business called to mind the excitement human babies express when tossed in the air and caught.

221

This was a happy time at the pond, and from what I could see, Lily was holding her own. Whatever it was that afflicted her did not seem to interfere with her ability to mother. One evening John and I spotted her heading our way with both her kits. The babies alternately swam by her side and rode on her back. When either or both fell behind, they paddled rapidly to catch up, then pawed their way up onto her tail or grabbed on to the ruff around her neck.

Midway across the pond, their excursion was interrupted by the sudden appearance of Buttercup. All four beavers greeted in a nose-to-nose configuration that resembled a slowly revolving weather vane. Then Buttercup departed, leaving Lily and the two kits equidistant from both shores. At first the kits started for the lodge, but Lily had another destination in mind. She seized a big lily leaf and held it in such a way that a part stuck out of each side of her mouth; with this she enticed her youngsters to swim alongside her, like two rabbits chasing a carrot. As luck would have it, she headed in our direction, and so we had a good long sighting of this unusual event. There was no doubt in either of our minds that Lily was deliberately holding the leaf just beyond the reach of the kits and that they were struggling to catch up with this tantalizing bait. In their eagerness to obtain it, the little ones surged forward and at times tried to mount Lily's back. Inevitably they slipped off, but they did not let that deter them. On they paddled to the farthest point they had yet traveled, all the while chasing a leaf.

When at last Lily relented and came to a halt, the three were not six feet from the bank where John and I were sitting. There she let the round lily pad fall from her mouth, and instantly both kits pounced on it. While they fed I got my first close-up view of them. They were as fat and sassy as if born to a young and healthy mother. And once again I had to thank Lily for coming through for me: the two were different shades of brown. I named the lighter one Sandy; the other Fern.

Once her new offspring had made it safely to the north shore and back, Lily became quite relaxed about leaving them and she took to visiting me again.

"Lily, I've ruined you," I told her. "You're not supposed to cotton up to people like this. It's dangerous."

But I knew that it no longer mattered what she did. Human beings would not bring her to grief, nature would. Lily was terminal; the feeble old beaver had little time left. And now she seemed so content to come out of the water and sit beside me and gnaw on whatever she found lying there that I had not the heart to deny her this small pleasure. And so I carried aspen branches to the pond and felt right about doing so. For where is it written that compassion must be renounced in favor of absolute detachment toward a wild animal whose days are numbered? And Lily hummed softly as she fed on my offerings.

One day I caught sight of her approaching my viewing station accompanied by Fern, the darker of her two kits. Quickly I laid out the aspen I had brought, then moved some distance away so that the baby would not learn to identify me with food. The two crawled out of the water, mounted the bank, and began to eat. I was impressed by how adept the baby was at rotating a branch, though she clearly seemed to prefer eating green leaves to stripping bark. Of even more interest was Lily's indulgence toward her kit. After nibbling the youngster's coat in a perfunctory effort to groom her, the two sat side-by-side and fed on the same branch in perfect harmony.

While I was watching this tender scene, who should come cruising down the shore but Huckleberry. And even as he swam back and forth, eyeing the pair on the bank, Dogwood and Daisy and Buttercup showed up. After some time Huckleberry mounted the bank and moved stealthily toward his mother and baby sibling, both of whom had their backs to him. Nevertheless, Lily was not caught off guard. With a hiss and mock thrust, she turned suddenly and sent him packing to water.

This setback did not, however, discourage Huckleberry, and soon Daisy summoned courage to take part in his raids. Silently, she trailed behind him as he stole up the bank, then, just as he pulled alongside Lily and was about to seize her branch, Daisy took hold of the downhill end of it and yanked it clear of both of them. Huckleberry was as startled as Lily by this unexpected turn of events, but even before he fully registered what had happened, Lily gave *him* a nip that sent him galloping to the pond.

This incident still did not put an end to Huckleberry's attempts

Lily grooms her kit Fern. / Then both feed side-by-side. (Note Lily's paralyzed paw.)

224

to steal Lily's branches. To a lesser degree, the other beavers also pressured the mother of them all to relinquish some of the aspen I brought her. Yet no beaver went on the attack or used force to obtain the coveted food. Nor did any respond in kind when Lily assertively defended what was rightfully hers. No real fights erupted. No blood was shed. Even the nips Lily laid on Huckleberry could not have been more than a pinch, for no mark was afterward visible. Not even a tuft of hair was lost. Given the kind of damage a beaver's teeth can do, Lily's bite mechanism must have been powerfully inhibited when she clamped down on him.

Thus I saw firsthand how fail-safe are the strategies employed by beaver families to prevent disputes from escalating into warfare. To have evolved such restraints, the species must have lived in close contact over eons of time. During such long ages, those individuals inclined to keep the peace survived longer and therefore produced greater numbers of offspring than did their more bellicose relatives, whose inclination to do one another in surely curtailed their breeding years. Eventually, the genes of the more socially adapted beavers swamped the population. That is how natural selection works. Despite a widely held notion to the contrary, the expression "the survival of the fittest" has little to do with battling for supremacy. Infighting, in fact, isn't conducive to species survival. What must be preserved, after all, are not a handful of belligerent *individuals*, but those *characteristics* that favor successful reproduction. Ultimately, it is the best mother, the best hunter, the best hider, the best builder, the best runner, and the most healthy and intact animal whose genes are passed to the next generation. And to remain healthy and intact, combat must be held to a minimum. Some species manage this by distancing themselves from others of their kind, each individual claiming and marking an exclusive living space for itself. The more socially evolved animals, however, having discovered the value of cooperation, had to invent complex strategies for getting along with their relatives—including some built-in inhibitions against fighting. Clearly the beaver is one of these.

Chapter Twenty-six

"The beavers have moved to Top Pond and they're felling trees that have trunks nearly two feet in diameter," I reported to John on the phone.

"Are you telling me that those lotus-eaters have finally decided to act like beavers and take down trees?"

"Oh come on, John, they've felled trees before now. Nothing like these, though. You should see the size of the six giants lying on the ground. And they're working on thirteen others—mostly oak, but some swamp maple and a large sour-gum."

John, who had recently traded country life in Massachusetts for a teaching career in New York City, was missing contact with the outdoors and quickly accepted my invitation to visit my colony. For my part, I could hardly wait to see his reaction to the transformation that had taken place, not only at Top Pond, but in the animals themselves.

These changes had occurred quite abruptly. In mid-September, I noted that not a single lily remained on Lily Pond. As a result, a new type of vegetation had begun to colonize the bottom muck, a green, gooey plant that did not look to be a food the beavers would find at all appetizing. Whether it was this transmutation of the pond ecosystem that prompted the beavers to relocate or an awareness that their winter fare had been entirely consumed, I cannot say, but that week I observed the yearling Daisy depart Lily Pond and I followed her. Over seven dams she climbed, then swam through the culvert that led under the road to Top Pond. There,

After moving to Top Pond, the Lily Pond colony begins felling trees in earnest.

227

for the first time, I saw what the beavers had been up to behind my back.

The place was humming with activity. The same lodge that had housed the Skipper, Second Mate, Sweet Potato, and Yam throughout the previous winter was now being enlarged and renovated by Buttercup, Huckleberry, and Dogwood for use by the Lily Pond colony. These three siblings, in fact, were so engrossed in what they were doing that they did not so much as cast a glance in my direction as I climbed a nearby hill to obtain a bird's-eye view of the busy scene. Each beaver was carrying out a different building task. Huckleberry repeatedly dived for bottom muck, which he then clasped between his chin and one forepaw and toted up the side of the teepee-shaped house, walking on three legs. Dogwood gathered leaves in his mouth, and this thatching material he easily transported up the steeply pitched roof on all fours. But it was Buttercup's approach to lodge renovation that most charmed me. Balancing loads of small branches on her forearms, she mounted the precipitous incline on her two hind legs, moving at an ever-accelerating rate so as to defy the pull of gravity. Upon arrival at some spot in need of attention, each beaver dumped his or her load and then pinned sticks, stuffed leaves, or slathered mud on to the existing framework of the house.

The result of all this uncoordinated yet mutually supportive labor was an island lodge of impressive dimensions. Already it stood five-and-one-half feet above the waterline and was ten feet long and twelve feet wide. Such a large shelter would provide more than adequate space to house all eight beavers, should the two-year-olds, Huckleberry and Buttercup, remain at home with the family for an extra winter. And indeed at this late date, it looked as if they would do just that.

But house restoration was only a part of the winter preparations taking place at Top Pond. Work on a food cache had already commenced. In a far corner of the pond, the Inspector General stood in shallow water and gnawed at the base of an oak tree that rose sixty feet into the air. A five-inch wedge on one side of its trunk suggested that the repetitious chomping I was hearing had for some hours served as rhythmic accompaniment to the other beavers' projects. Now the scratching of teeth on wood stopped quite suddenly

Buttercup mounts the lodge bipedally, toting building material on her forelegs.

as Daisy joined the big beaver at the tree. It was obvious that the Inspector General did not welcome any assistance, for as Daisy raised up on her hind legs to test the rough bark with her nose, he demonstrated his annoyance with a show of temper. First he attacked the tree, clawing at it fitfully. Then he hissed. And finally he took off in a huff, leaving the job of felling the giant oak entirely to the yearling.

This ability to refrain from acting out a hostile impulse that has been aroused by mate or kin, instead directing it at some inanimate object (or even self), is called "displacement behavior" by psychologists. It is seen in many social species, including man (socking a wall instead of a wife). I once observed it in a pet coyote who, when reprimanded, inflicted bites on his own hind leg rather than attack his keeper (his pack leader). Now I saw that the strategy was also part of the beaver's behavioral repertoire, though I had never heard it reported. Still, I was not surprised, given how socially evolved is the species.

Over the next two weeks, I watched the Inspector General, Huckleberry, Buttercup, Dogwood and Daisy convert Top Pond to their liking. Meanwhile, Lily and her two kits continued to live at Lily Pond. During this period the old matriarch had progressively deteriorated. In addition to being bedraggled and unstable on her feet, she now seemed spaced-out. I feared she would not last much longer and every night, to reassure myself that she was still alive, I visited Lily Pond before climbing the hill to watch what was transpiring at Top Pond.

"Some beaver must be making trips to Lily Pond to groom the kits," I told John the weekend he came up. "I doubt Lily is capable of doing it, yet they're sleek as seals. I wish whoever it is would tend to Lily's coat too. Her fur is a mess. She's wet all the time, even after she's been on land for an hour."

Normally, a mother beaver and her kits engage in frequent mutual grooming sessions, a practice that helps socialize the youngsters. Now, however, Lily would have great difficulty twisting herself into such embraces as mutual grooming demands. And with only one functional front paw, she no longer made any effort to comb even herself. As a result, her heavy undercoat had become so sopping wet that I had little hope of it ever drying out.

"She's in bad shape, the worst I've seen yet," I explained to John as we headed for the pond. "If she lives, I expect she and her kits will winter alone at Lily Pond. I don't believe she is capable of hauling herself up the hill and over seven dams to join the rest of the colony at their new winter quarters."

John shot me a look that told me his mind had made a leap forward in time and was struggling with the question I didn't want to voice—that of winter food.

"There are the kits to think about too," was all I said.

We had reached that point on the path where Lily Pond loomed into view, and from there, we watched the wake of a beaver, a rippling arrowhead moving across the water toward the old and by now dilapidated lodge. I raised my binoculars to identify the beaver at the point of all that backwash just as he tipped up and dove into the old homestead.

"That isn't Lily," I said. "She veers to the right. I think it's the Inspector General, although it seems kind of strange to see him back in these parts again."

Just seconds after disappearing, the big beaver—who did indeed turn out to be the Inspector General—surfaced outside the lodge with a kit on either side of him. Immediately, all three took off for the marshy inlet—the father beaver setting such a no-nonsense pace that the kits had trouble keeping up with him. Even though they paddled like demons, from time to time one or the other was forced to grab hold of the big male's ruff or ride his tail so as not to be left behind.

Just so, parent and kits covered the distance between the lodge and Square Pond in a record eight minutes, whereupon all three scaled Square Pond's high dam, crossed its waters and pushed on to the next higher pool. There again they crawled over the backside of a dam, dragging their flat tails behind them, and on they traveled. Each pool they exited was left rocking with excitement. Each pool they slipped into convulsed into expanding rings.

"He's escorting the kits to Top Pond, and we're seeing it happen," I marveled. "Think if we'd arrived five minutes later. We would have missed all this."

Midway up the chain of beaver-made pools that ascended the slope like terraced rice paddies, the Inspector General paused. One of the two kits was no longer at his side. The big beaver and the baby still in his company waited. From where John and I stood, we could see the laggard, Sandy, struggling to make it up and over the backside of the steepest dam. After sliding backward three times, he finally cleared the hurdle and raced to join the others. Immediately they all set off again, full steam ahead.

"If I had any lingering doubts about what is going on here, they have now been put to rest," John said. "Clearly, the Inspector General is *purposefully* bringing those kits to Top Pond."

When the three had made it over the last dam and were entering the road culvert that opened onto Top Pond, we ran across Welch Drive, arriving just in time to see father and kits emerge. Now that the youngsters had been brought into safe harbor, the Inspector General relaxed, floated on the water and did nothing at all for a few minutes. Then he slowly made his way to a partially cut tree and resumed work on it. The entire trip had lasted twenty minutes.

After that Lily was all alone. Though I was glad the babies would not have to exist on what would undoubtedly be paltry winter fare, I was sorry the old matriarch was now denied the solace of their presences. I felt certain she missed them, for she took to visiting

The Inspector General returns to Lily Pond to fetch his kits, then escorts them over seven dams to Top Pond, where the colony has relocated.

my viewing station and gazing at my face in a way that reminded me of an experience I once had with a wild mustang—a young male who had been run out of his father's harem. For want of equine companionship, the two-year-old colt began following me around. I named him Lonesome and soon discovered that he enjoyed the sound of my voice. Now I could hardly bear to see Lily so bereft and Lily Pond so deserted, and so I came early and left aspen branches onshore for the old girl, then went off to Top Pond before she woke up.

Meanwhile the mud-slathered walls of the Top Pond lodge had become as solid as adobe and more than a foot thick. And the beavers' food cache was stuffed to capacity with alder poles and at least nineteen sizeable trees the beavers had felled and cut up. Still the animals continued to labor. Three more huge trees were brought down in such a way that their crowns landed precisely on top of the food raft.* I no longer worried that the two-year-olds, by stay-

---

*The beaver isn't supposed to be able to control the direction a tree falls; these must have been fortuitous drops.

ing on, would put a strain on the food budget. On the contrary, it was they who had laid in the lion's share of what had been stockpiled—more than enough to keep the entire colony alive, however severe the winter.

One night, two-and-one-half weeks after the kits were escorted to Top Pond, I spotted Lily there. For a moment I thought I was seeing things. But no, the decrepit animal I was looking at was indeed she.

"You made it!" I shrieked.

My outburst sent her underwater. When she surfaced, however, she seemed to recognize me, for she nonchalantly swam back to the branch she had just stolen from the food cache and began gnawing on it. I could not imagine how she had hoisted her half-paralyzed body over seven dams to join the colony, but in so doing, she certainly put my mind at ease. Whether or not she lasted through winter, now at least she would spend what time remained to her in the company of her mate and offspring.

Chapter Twenty-seven

L
ily lived for twenty-three days after I spotted her at Top
Pond, days of balmy weather and untroubled ease. With so
many partially cut-up trees lying on their sides, food was
readily available to her. Moreover, she rarely lacked company; the
kits, Fern and Sandy, seldom left her side. Even the yearlings,
Dogwood and Daisy, frequently towed their food branches to wherever their mother and younger siblings were dining and joined the
party.

It was during one of these beaver picnics that I observed a piece
of business I had heard reported but never seen, that is, dancing.
One evening Lily and the kit Fern were sitting side-by-side in the
water debarking branches when suddenly Dogwood bobbed up and
sought a place near them. Fern objected, at first only vocally, but
then with a vigorous thrust of her body. Momentarily, this mock
attack startled the yearling, and he swam away. Before long, however, he circled back and tried again to gain admittance to the
company of his mother and sister. No soap! Fern sent him packing
with another lunge (the kind of sudden jump one automatically
performs when shouting "boo!"). But Dogwood was not discouraged. He made a third try. And this time he did not back away
when the kit tried to spook him, but held his ground, sat up on
his haunches and began to dance. For some seconds he shimmied
and shook and threw his head around like a burlesque queen. I had
to laugh to see the love handles and jelly rolls ripple up and down
his fat body.

The next day I looked through the beaver literature for information on dancing. Lars Wilsson described it as a display, or body language, used by one beaver to invoke his dominance over another. According to him, the more subordinate of the two animals, upon witnessing such behavior in the other, often responds by grooming the dancer as an act of appeasement.

Having seen this behavior only once, I have little to add to Wilsson's explanation except to note that Dogwood's hula hula did put an end to Fern's rejection of him and gained him a place beside her and Lily. Moreover, having long been aware of the importance that pecking order plays in the lives of many social species, I was pleased to witness some evidence that a dominance hierarchy existed within my beaver colony. For in those animals that have evolved such stratified societies, the blood of kin is rarely shed. When rank has been settled ahead of time, the more dominant animal need only posture before a subordinate one to remind him of some past defeat, and combat is avoided.

Even so, it was hard for me to work out just who was dominant over whom in this family of beavers I had come to know so well. As far as I could see, age conferred no special status. True, infant kits enjoyed catering service, but certainly not because they were dominant! As soon as the kits became relatively self-sufficient, not only were they treated like every other beaver, but they behaved like them too. Half-grown kits did not hesitate to swipe branches from their elders, no matter how many mock thrusts and vocal chastisements were directed at them. By the same token, adult beavers were not above nipping branches from their juniors, and they, too, ignored what verbal abuse and sham attacks their actions aroused. Nor did I notice that one sex was dominant over the other. Perhaps the animals deferred to rank within the confines of the lodge, but inasmuch as I was not privy to transactions that took place behind the thick walls of their moat-protected fortress, I was never able to ascertain who, if anyone, lorded it over whom.*

*In a study by Harry Hodgdon and Joseph Larson, published in *Animal Behavior* in 1973, the adult female is reported to be the dominant animal. After many more years of beaver study, however, Hodgdon disavowed this idea, having later found no consistent pattern of dominance by either sex. Nevertheless, his early paper continues to be quoted in popular literature. Still another study of beaver dominance was undertaken by P.E. Busher in California's Sierra Nevadas. This study suggests an age-related hierarchy, older beavers being dominant over younger ones.

Lily makes a feeble attempt to gnaw a tree, while her kits look on attentively.

By now the days were short and I was forced to watch the animals almost entirely by lantern light. One evening in early November, my beam caught Lily gnawing on a partially cut sour-gum tree that grew beside the water. Even in the dark the kits accompanied her everywhere, and now they looked on, wide-eyed, as she pulled wood chips from the tree's trunk. This was Lily's last attack on a tree, and I was touched by the youngsters' keen interest in what she was doing. When they were not fingering and nibbling the chips she dropped, they sat up on their haunches and watched her work. Though conditions were poor, I photographed the scene, for the kits' attentiveness lent support to my belief that young beavers gain valuable experience simply by observing their elders.

On Veterans Day snow fell thick and fast and blotted out all landmarks and horizons, and softened the sharp outlines of trees and beaver lodges, and frosted the slushy ice that had formed on the pond. Dan and John and I drove to the park in Dan's four-wheel-drive van, much to the consternation of a park patrolman, who rolled his eyes and asked if we had lost our senses. It was a freak storm. Soon bright weather would return, bringing with it some confused migratory birds. Winter, after all, does not begin all at once on November 11 at 41 degrees latitude.

"Well the beavers are ready for this bad weather, in any case," I commented to Dan. "Wait until you see their food cache."

I expected most of the colony to be out and working to push aside the heavy slush that had formed on their pond before it turned to hard ice and imprisoned them. And indeed they were. Their coats, winter-thick, were so perfectly groomed and oiled that not a drop of water penetrated their outer guard hairs as they sloshed about in the frigid water.

"I only hope that Lily remains inside the lodge tonight," John said. "You haven't seen her for some time, Dan. If she does comes out, I'm afraid you'll be in for a shock. She can't groom herself anymore and she's a mess."

"If she comes out tonight, she's doomed," I added. "She's soaked to the skin all the time. If she comes out tonight, she will surely die of exposure."

We wandered about the edge of the pond, each one of us captivated by a different one of the seven beavers who were working, independently, to carve pathways through the mush. As usual, they pushed down on the soft ice with their black satiny forepaws, then went below and bumped away globs of it with their rounded backs. Sandy and Fern seemed to enjoy this new game.

Watching the efforts of the two youngsters evoked memories of Blossom and Lotus and set me to reminiscing. Whatever had become of those first little beavers, those fall-born kits I had met in the dark of night more than three years earlier? Had they found a suitable site to colonize? And where were the Skipper and Second Mate and Sweet Potato and Yam? One would think that animals who put so much work into their habitats would be reluctant to abandon them. But now even the Lily Pond beavers were laboring in a less lovely site than the one that they had given rise to many years earlier. Did this strange pond-making animal not belong to its own creation? Did its creation not belong to it? How many work years had been invested in Lily Pond? How much dredging and felling and damming and lodge construction? How much maintenance and scent marking and kit production had gone on there? Yet the colony had allowed their claim to lapse, had moved out en masse to pioneer a new site.

What a lesson I could learn from *Castor canadensis*. It was past time for me to do something similar, to end my beaver study and

Beavers try to forestall winter incarceration by bumping through ice cover and pressing on the edge of the opening to break off more ice slabs.

238

seek a new challenge. Yet I suffered separation pangs just thinking about saying goodbye to Lily Pond and all the beavers who had lived there. Why was I not able to close the book on this experience with the same freedom of spirit exhibited by my animal subjects? How had I become so attached to a place and its animal inhabitants that I was incapable of moving on?

While thinking these thoughts, I spotted Lily. To my horror, she was making an effort to keep the waterways open. Like the others, she was bumping ice and diving, bumping ice and diving.

"Oh Lily," I could barely speak. "Oh Lily, you should stay in the lodge in this weather."

She didn't hear me. She seemed unaware of my presence.

"Oh Lily."

I watched as she sank under the slush and disappeared. I waited, but she did not surface. I listened, but the pond was as silent as the snowflakes settling on my eyelashes.

"Oh Lily, is this how you leave us?" I asked at last.

Slowly, painfully, she rose up through the ice, but it was clear she had not the strength to shake the slush from her coat. She was past breaking pathways. She was past mothering kits, past felling trees, past building lodges, past stopping streams. She lay on the mush, unseeing, her transparent eyelids closed. Then with great effort she forced her frail body under the ice cover and did not come up again. I stared at the spot until snow covered it over, but I knew I had seen Lily for the last time.

I did not seek out the others to tell them what had happened. Lily was my special beaver, and the others were somewhere else, watching something else. They might not say the right thing. They might even try to console me by denying she was gone. I was tempted to wade out and gather her in my arms, to hold her as I had so often wanted to do; but instead I stood on the shore and thought about her life and about how she had gone out with her boots on, so to speak, the way a beaver ought to leave.

In the dark, the stark forms of the other beavers continued to move about like Arctic icebreakers, still visible against the white snow. It was a beautiful sight, but all I could think of was how painful it is to give up what you love, even when it has already changed into something else.

After a while I heard John call. The roads were bad and it was time to go, he said.

He was right, of course. But I wasn't quite ready yet. My leave-taking would have to be more gradual. For the time being, I would continue to visit the colony every night. Soon, however, winter would seal the animals away and create a natural break in my viewing routine. I would visit them in the spring, of course. I would stop by the pond to assure myself that all the animals had made it through the cold season. And later on in summer I would pay a call on the family to learn if the Inspector General had found a new mate and to meet any new kits he might have sired. My leave-taking would have to be gradual. After all, I couldn't be expected to tear up roots with the casualness of a beaver.

John called again, and this time I answered him.

"I'm coming, right away."

But I took one more minute to break off a branch and lay it on the shore.

"That's for you, Lily," I whispered. "Godspeed, wherever you are."

Then I took one long look around at all the others.

"So long, gnomes," I called softly to them. "I'll be seeing you."

Postscript

Soon after ice-out in 1988 I succeeded in getting a full count of the seven beavers who had wintered at Top Pond. All looked healthy and well fed, despite the length and severity of the cold season. Four members of the family—Huckleberry, Buttercup, Daisy, and Dogwood—were due to emigrate. Huckleberry's and Buttercup's departures were, in fact, overdue by a year. Nevertheless, while I watched and waited for these animals to take their leave, I was aware that any one of the four might pair up with the widowed Inspector General and remain at the pond. None did, however. On the contrary, the Inspector General himself departed and was the first to leave. Did he go in search of a new mate?

Soon thereafter Huckleberry and Buttercup disappeared, and a bit later, Daisy vanished from the scene. That left Daisy's litter-mate, Dogwood, as heir-apparent to the colony's extensive water-works and as sole role-model to the previous year's kits, Fern and Sandy. Perhaps this peculiar situation explains why that two-year-old failed to hear, or at least paid no heed to, the call to emigrate.

That spring he and the two youngsters wandered up and down the seven terraced pools that connected Top Pond and Lily Pond and fed on sedges and ferns, growing along their banks. Then unexpectedly, I discovered that the threesome had been joined by a mature beaver of unknown origin. I theorized that the newcomer was a long-departed family member who, having fallen upon hard times, had returned home. And when I caught sight of a split in his tail I felt sure that he was none other than Blossom.

Fern feeds on ferns growing alongside the beaver-made pools that connect Top Pond and New Pond.

By then Blossom would have been a few weeks short of four and thus would not have known Fern or Sandy, who were born after his departure. No doubt he had known Dogwood as a kit, however, and Dogwood must have recognized him, for the new settler was immediately accepted on the property.

I was glad I had not come upon such an odd assortment of animals when I began my study, for had I done so, I would never have been able to make sense of them. There was no mated pair among them, though all behaved amicably toward one another and even engaged in mutual-grooming sessions. Still, they did not hole up together to sleep. Dogwood, who spent most of his waking hours at Lily Pond, took over the original family lodge there, which by then was in quite a dilapidated state. Fern, too, moved back to Lily Pond, but slept her days away in the lodge by the marsh where she had been born. Sandy continued to make use of the Top Pond lodge until a family of otters took possession of it in July. I never figured out where Blossom slept.

Of course no kits arrived that year to create a kind of center-of-focus for the assorted siblings. Nevertheless, they achieved real cohesion when, in late summer, all four began to work together to fell a large yellow birch in back of Square Pond. Shortly thereafter, they set about renovating the old family lodge, which was still being occupied by Dogwood. After that, I found several more cut trees about the property and discovered that the beavers had begun caching branches. Obviously, they were preparing to winter together at Lily Pond.

Another development occurred in the fall, one that greatly surprised me. In September, bullhead and fragrant water lilies began to dot Lily Pond. For more than a year, not a single pad had surfaced, and now the sight of these aquatic plants growing in patches here and there gave rise to hope that the good old days were about to return.

But what was going on? Water lilies surfacing in September?

I speculated that the plants had arisen from seeds, which had been lying fallow in the bottom muck, and that it had taken this many months for them to germinate.

Normally, water lilies arise from extensive root systems, which continue to grow and send up multiple plants from early spring until the first frost of autumn. But like all tuberous species (such as the day lily and the iris), water lilies also make seeds, which can and do sprout under special conditions. Nevertheless, their primary means of reproduction is through their ever-spreading root systems. Now, however, since the beavers had cleaned out the pond of lily rhizomes, these new plants, it appeared to me, had to have sprouted from seeds. Likely these late-blooming upstarts were being supported by small roots, which would need time to collect solar energy and spread. If lily rhizomes were ever again to become a viable winter food for the beavers, the animals would have to refrain from eating them during the oncoming cold season.

But would they?

It is not the business of *Castor canadensis* to preserve its food resources and so extend its tenure in a place. On the contrary, it is the business of the beaver to create conditions that eventually force it to leave home. Sooner or later Lily Pond would be abandoned, and the animals I knew (or their descendents) would set off to revitalize the landscape elsewhere; for the beaver is truly our best agent for renewal, our foremost conservationist.

Still, I couldn't help but feel joy at the sight of lilies on the pond again, and I couldn't help but hope that the colony's lengthy sojourn at Top Pond had disposed them toward a diet of bark. And so once again that wellspring of optimism, which energizes most human endeavors, started percolating in me. Lily Pond certainly seemed to be resilient. And the beavers did not appear to be finished with it yet.

Hope Ryden
Fall 1988

What Happened After That

T he other day it struck me how much time has passed since I concluded my study at Lily Pond in 1988. Although I have continued to visit the place, it has been considerably more difficult for me to do so since John and I married in 1989 and bought a place in the Catskills. That's when I gave up my nearby cabin, which had long afforded me quick access to the Lily Pond beavers—an easy seven-minute drive, a one-minute walk to my viewing post, and (presto!) the late, late show would begin!

Today, a trip to Lily Pond requires an hour and a half drive each way, and my visits are not so frequent that I can learn to distinguish one inhabitant from another, and identify their unique characteristics. I miss that. Still I make the trip as often as I can, and Lily Pond continues to bestow gifts on me—the discovery of a rare wildflower, the perfect place to find and photograph insects of my children's book on crawlers and flyers, sightings and soundings of weasels and whippoorwills, evidence of new beaverworks, and, of course, the beauty of the pond, itself. The best news is that Lily Pond is flourishing and once again covered—literally matted—with lilies.

None of this would be happening if beavers had not remained in residence at some low level to maintain their dam. Granted their numbers have been minimal. Since 1989, never more than three beavers, and usually only two, have occupied the place. What's more, not a single kit has shown up since Fern and Sandy emerged from the lodge in the spring of 1987. Yet this is all to the good. While beaver numbers have been low, the lily crop has had a chance to make a full recovery. At the same time, a skeleton crew of beavers has kept the dam in good repair and thus provided the recovering lilies with the water they must have to live.

As I see it, the stage is now set for beavers once again to reproduce. Perhaps there will be a series of annual births, as happened when I first came upon this lily-covered pond some thirteen years ago. During the five-year period of reproductive vigor that followed, I never guessed that I had entered the picture at the very onset of a cycle, one that would end precisely when my study was completed. What luck to have been in just the right place at just the right time so as to witness the many family dramas that unfolded before my eyes! What if I had entered the scene at the end of this cycle instead of at its beginning?

But maybe luck is not the right word to describe such fortuitous timing. Luck is cheap, and Lily Pond has been so generous to me I want a better word—one that acknowledges the benevolence of the place and the gratitude I feel for what it brought me. No, luck doesn't say it. The word I am searching for is blessed. I was blessed.

Hope Ryden
1996

Appendix

However beneficial the beaver, the animal is viewed as a nuisance in certain quarters for the reason that it does not take man's projects into account when it builds its dams. Frequently it plugs culverts and thus floods roads, fields, and railbeds. Tearing out these unwanted dams has little effect, for beavers are not easily discouraged and will quickly rebuild them. Placing a steel grate against a culvert opening doesn't work either. It merely provides a backing against which beavers can build. As a result, it is the practice of most state fish-and-game officials to respond to complaints of beaver flooding by trapping and killing the offending animals. Even this solution, however, is a short-term one. Where water flows, new beavers will soon appear. Nature hates a vacuum.

Now, however, someone has come up with an answer. Canadian oil and gas consultant Neil Thurber has designed a "beaver baffler" that works, a simple device consisting of a long mesh tube, three layers thick. One end is inserted a short way into the culvert opening so that most of the length of the tube extends some distance downstream. When properly fitted, water spurts through this undammable mesh device in so many places that beavers are at a loss to stop it.

Specifically, the cylindrical core of this device is made of concrete reinforcing mesh and around this tube is wrapped a second layer of one-inch-by-one-inch mesh fencing. A third layer, consisting of reinforcing mesh, encloses both of these. Then all three layers are wired so that they are held apart at six-inch distances. Finally, the

upstream end of the device is capped with wire to prevent
beavers from entering the tube and damming it from the inside.

A question remains as to how far the tube need extend beyond
the culvert. Hope Buyukmihci at her beaver refuge in the New
Jersey Pine Barrens has made tests using varying lengths, and her
results suggest that shorter lengths than that prescribed by
Thurber can sometimes be effective.

Photo of Hope's Tests

*The tube labeled "Graff Baffler" extends fifteen feet beyond the culvert (length recommended by
Thurber). The tube labeled "Township Stop" is only five feet long. Hope Buyukmihci found
both bafflers to be effective.*

  For more information, write to:

Hope Buyukmihci   *or*   Sharon Brown
Unexpected Wildlife Refuge   Beavers, Wetlands and Wildlife
Newfield, NJ 08344   P.O. Box 591
                                        Little Falls, NY 13365

Sources

Aeschbacher, A., and G. Pilleri. "Observations of the Building Behavior of the Canadian Beaver in Captivity." *Investigations on Beavers*, vol. 1, pp. 91–98 Berne, Switzerland: Institute of Brain Anatomy, 1983.

Aspen: Dancer on the Wind, documentary film co-produced by the British Broadcasting Corporation and WNET-TV, New York City. Telecast on WNET on April 6, 1986.

Bollinger, K. S.; H. E. Hodgdon; and J. J. Kennelly. "Factors Affecting Weight and Volume of Castor and Anal Glands of Beavers." *Acta Zool. Fennica*, 174, 1983, pp. 115–116.

Brody, Jane. "Return of the Beaver Yields New Opportunity for Study of a Builder." *The New York Times*, August 25, 1987.

———. "Surviving in the Cold: Strategies Revealed in New Studies." *The New York Times*, December 15, 1987, pp. C-1 and C-8.

Bush, G. L.; S. M. Case; A. C. Wilson; and J. L. Patton. "Rapid Speciation and Chromosomal Evolution in Mammals." *Proceedings of the National Academy of Science, U.S.A.*, vol. 74, no. 9, September 1977, pp. 3942–3946.

Busher, P. E. "Interactions Between Beavers in a Montane Population of California." *Acta Zool. Fennica*, 174, 1983, pp. 109–110.

Buyukmihci, Hope Sawyer. *Hour of the Beaver*. New York: Rand McNally, 1963.

———. *Beaver Defenders Newsletter*(s), personal letters and discussions.

Champagne, Glenn C. "The Beaver in New York." *The Conservationist*, New York Department of Environmental Conservation bimonthly publication, August/September 1971, pp. 18–21.

Courcelles, R., and R. Nault. "Beaver Programs in the James Bay Area, Quebec, Canada." *Acta Zool. Fennica*, 174, 1983, pp. 129–131.

Danilov, P. I., and V. Ya. Kan'shiev. "The State of Populations and Ecological Characteristics of European and Canadian Beavers in the Northwestern U.S.S.R." *Acta Zool. Fennica* 174, 1983, pp. 109–110.

Doboszynska, T., and W. Zurowski. "Reproduction of the European Beaver." *Acta Zool. Fennica*, 174, 1983, pp. 123–126.

Grey Owl. *Pilgrims of the Wild*. New York: Charles Scribner's Sons, 1935.

Hodgdon, Harry Edward. "Social Dynamics and Behavior Within an Unexploited Beaver Population." PhD thesis, University of Massachusetts, August 1978.

Hodgdon, H. E. and R. A. Lancia. "Behavior of the North American Beaver." *Acta Zool. Fennica* 174, 1983, pp. 99–103.

————— and J. S. Larson. "Some Sexual Differences in Behavior within a Colony of Marked Beavers." *Animal Behavior*, 21, February 1, 1973, pp. 147–152.

Irving, Washington, *Astoria*, vol. 1. Philadelphia: Carey, Lee and Blanchard, 1836, pp. 38–40.

Lancia, Richard A.; Wendell E. Dodge; and Joseph S. Larson. "Winter Activity Patterns of Two Radio-Marked Beaver Colonies." *Journal of Mammalogy*, 63(4), 1982, pp. 598–606.

————— and H. E. Hodgdon. "Observations on the Ontogeny of Behavior of Hand-Reared Beavers." *Acta Zool. Fennica* 174, 1983, pp. 117–119.

Larson, Joseph S. "Age Structure and Sexual Maturity Within a Western Maryland Beaver Population." *Journal of Mammalogy*, vol. 48, no. 3, August 21, 1967, pp. 408–413.

————— and J. R. Gunson. "Status of the Beaver in North America." *Acta Zool. Fennica* 174, 1983, pp. 91–93.

————— and S. J. Knapp. "Sexual Dimorphism in Beaver Neutrophils." *Journal of Mammalogy*, vol. 52, no. 1, February 1971, pp. 212–215.

Lavrov, L. S. "Evolutionary Development of the Genus Castor and Taxonomy of the Contemporary Beavers of Eurasia." *Acta Zool. Fennica* 174, 1983, pp. 87–90.

Leighton, Alexander H. "Notes on the Beaver's Individuality and Mental Characteristics." *Journal of Mammalogy*, vol. 13 (2), May 1932, pp. 117–126.

————. "Notes on the Relations of Beavers to One Another and to the Muskrat." *Journal of Mammalogy*, vol. 14 (1), February 1933, pp. 27–35.

Leyhausen, Paul. *Cat Behavior: The Predatory and Social Behavior of Domestic and Wild Cats*. New York and London: Garland S.T.P.M. Press, 1979, p. 138.

McNamara, Robert J. "An Inside Story." *The Conservationist*, New York Department of Environmental Conservation bimonthly publication, January/February 1987, pp. 27–30.

Miller, John W. "Return of the Beaver." *Natural History*, June 1972, pp. 67–72.

————, "Knowing Ice." *Audubon*, January 1980, pp. 35–39.

Mills, Enos A., *In Beaver World*. New York: Houghton Mifflin Co., 1913.

Morgon, Lewis H., *The American Beaver and His Works*. Philadelphia: J. B. Lippincott and Co., 1868.

Mosher, Nancy D. "The Floristic and Successional Development of Three Abandoned Beaver Meadows in Western Massachusetts." Master of science thesis, Smith College, May 1981.

Muller-Schwarz, D.; S. Heckman; and B. Stagge. "Behavior of Free-Ranging Beaver at Scent Marks." *Acta Zool. Feneca* 174, 1983, pp. 111–113.

Olsen, Glenn H. "Beaver Temperature Monitered by Radio-Telemetry." Master of Science thesis, University of Massachusetts, February 1980.

Owen, Roger C.; James J. F. Deetz; and Anthony D. Fisher, eds. *The North American Indians: A Sourcebook*. New York: The MacMillan Company, 1967.

Patenaude, Françoise. "Care of the Young in a Family of Wild Beavers." *Acta Zool. Fennica* 174, 1983, pp. 121–122.

————, "Une Année Dans La Vie Du Castor." *Les Carnets de Zoologie*, vol. 42, no. 1, 1982, pp. 5–12.

Pilleri, G. "Ingenious Tool Use by the Canadian Beaver in Captivity." *Investigations on Beavers*, vol. 1, pp. 99–100. Berne, Switzerland: Institute of Brain Anatomy, 1983.

————, "Nervous System of Castor Canadensis." *Investigations on Beavers*, vol. 1, pp. 19–59. Berne, Switzerland: Institute of Brain Anatomy, 1983.

Radford, Harry V. "History of the Adirondack Beaver: Its Former Abundance, Practical Extermination and Reintroduction." *Report of the [New York State] Forest Fish and Game Commissioner*, 1908.

Richard, P. B. "Mechanisms and Adaptations in the Constructive Behavior of the Beaver." *Acta Zool. Fennica* 174, 1983, pp. 105–108.

Richards, Dorothy, with Hope Sawyer Buyukmihci. *Beaversprite: My Years Building an Animal Sanctuary*. San Francisco: Chronicle Books, 1977.

——— conversations, 1975.

Ryden, Hope E. "Let's Hear It for the Eager Beaver." *The New York Times Magazine*, December 15, 1974, pp. 38–46.

Shields, William M. "Genetic Considerations in the Management of the Wolf and Other Large Vertebrates: An Alternate View." *Proceedings of the Wolf Symposium*, Edmonton, Alberta, Canada, May 12–14, 1981, pp. 90–92.

Stadler, Sandi. "Roadkills: Reducing the Death Rate." *The Animal's Agenda*, vol. 7, no. 8, October 1987, pp. 32–37.

Theberge, John. "Considerations in Wolf Management related to Genetic Variability and Adaptive Change." *Proceedings of the Wolf Symposium*, Edmonton, Alberta, Canada, May 12–14, 1981, pp. 86–89.

Warren, Edward Royal. *The Beaver: Its Works and Its Ways*. Baltimore: The Williams and Wilkins Co., 1927.

Wilson, A. C.; G. L. Bush; S. M. Case; and M. C. King. "Social Structuring of Mammalion Populations and Rate of Chromosomal Evolution." *Proceedings of the National Acadamy of Science*, vol. 72, no. 12, December 1975, pp. 5061–5065.

Wilsson, Lars. *My Beaver Colony*. New York: Doubleday, 1968.

———, "Observations and Experiments on the Ethology of the European Beaver." *Viltrevy*, vol. 8, no. 3, 1971, pp. 115–266.

Wise, Helen M. "Department of Forestry and Wildlife Reaps Wide Acclaim for its Beaver Studies: Family Affairs, Population Control and Twenty-Seven Hour Days." *The Alumnus*, University of Massachusetts at Amherst, August/September 1983, pp. 10–11.

Zurowski, W. "Worldwide Beaver Symposium, Helskinki 1982: Opening Remarks." *Acta Zool. Fennica* 174, 1983, pp. 85–86.

Index